Martin Taranetz

System-Level Modeling and Evaluation of Heterogeneous Mobile Networks

Martin Taranetz

System-Level Modeling and Evaluation of Heterogeneous Mobile Networks

The Symbiosis of Theory and Simulations

Südwestdeutscher Verlag für Hochschulschriften

Impressum / Imprint

Bibliografische Information der Deutschen Nationalbibliothek: Die Deutsche Nationalbibliothek verzeichnet diese Publikation in der Deutschen Nationalbibliografie; detaillierte bibliografische Daten sind im Internet über http://dnb.d-nb.de abrufbar.

Alle in diesem Buch genannten Marken und Produktnamen unterliegen warenzeichen-, marken- oder patentrechtlichem Schutz bzw. sind Warenzeichen oder eingetragene Warenzeichen der jeweiligen Inhaber. Die Wiedergabe von Marken, Produktnamen, Gebrauchsnamen, Handelsnamen, Warenbezeichnungen u.s.w. in diesem Werk berechtigt auch ohne besondere Kennzeichnung nicht zu der Annahme, dass solche Namen im Sinne der Warenzeichen- und Markenschutzgesetzgebung als frei zu betrachten wären und daher von jedermann benutzt werden dürften.

Bibliographic information published by the Deutsche Nationalbibliothek: The Deutsche Nationalbibliothek lists this publication in the Deutsche Nationalbibliografie; detailed bibliographic data are available in the Internet at http://dnb.d-nb.de.

Any brand names and product names mentioned in this book are subject to trademark, brand or patent protection and are trademarks or registered trademarks of their respective holders. The use of brand names, product names, common names, trade names, product descriptions etc. even without a particular marking in this work is in no way to be construed to mean that such names may be regarded as unrestricted in respect of trademark and brand protection legislation and could thus be used by anyone.

Coverbild / Cover image: www.ingimage.com

Verlag / Publisher:
Südwestdeutscher Verlag für Hochschulschriften
ist ein Imprint der / is a trademark of
OmniScriptum GmbH & Co. KG
Heinrich-Böcking-Str. 6-8, 66121 Saarbrücken, Deutschland / Germany
Email: info@svh-verlag.de

Herstellung: siehe letzte Seite /
Printed at: see last page
ISBN: 978-3-8381-5137-3

Zugl. / Approved by: Wien, TU, Diss., 2015

Copyright © 2015 OmniScriptum GmbH & Co. KG
Alle Rechte vorbehalten. / All rights reserved. Saarbrücken 2015

Abstract

The cumulative impact of co-channel interferers, commonly referred to as aggregate network interference, is one of the main performance limiting factors in today's mobile cellular networks. Thus, its careful statistical description is decisive for system analysis and design. A system model for interference analysis is required to capture essential network variation effects, such as base station deployment- and signal propagation characteristics. Furthermore it should be simple and tractable so as to enable first-order insights on design fundamentals and rapid exchange of new ideas. Interference modeling has posed a challenge ever since the establishment of traditional macro-cellular deployments. The recent emergence of heterogeneous network topologies complicates matters by contesting many established aspects of time-honored approaches. This thesis presents user-centric system models that enable to investigate scenarios with an asymmetric interference impact.

The first approach simplifies the interference analysis in a hexagonal grid setup by distributing the power of the interfering base stations uniformly along a circle. Aggregate interference is modeled by a single Gamma random variable. Its shape- and scale parameter are determined by the network geometry and the fading. The second model extends the circular concept by non-uniform power profiles along the circles. It enables to map substantially large heterogeneous out-of-cell interferer deployments on a well-defined circular structure of nodes. Thereby it considerably reduces complexity while preserving the original interference statistics. The model is complemented by a new finite sum representation for the sum of Gamma random variables with integer-valued shape parameter that allows to identify candidate base stations for user-centric base station collaboration schemes as well as to predict the corresponding rate performance. The third approach applies stochastic geometry to model two-tier heterogeneous cellular networks with respect to the topology of an urban environment. It tackles the asymmetric interference impact by a virtual building approximation and introduces a new signal propagation model that directly relates to the topology characteristics such as building density and -size, which can straightforwardly be extracted from real world data.

In the last part of the thesis, the applicability of the introduced models is validated against simulations with the Vienna LTE-Advanced Downlink System Level Simulator. For this purpose, the analytical models are calibrated against results from LTE-Advanced link level simulations. This part also complements the hitherto user-centric investigations with a system-wide performance evaluation, addressing the impact of user clustering as well as small cell density- and isolation. Particular focus is laid on a systematic and reproducible simulation methodology as well as appropriate performance metrics, since conventional figures of merit tend to conceal performance imbalances among users.

Kurzfassung

Der kumulative Einfluss von Gleichkanalstörern - häufig auch als aggregierte Interferenz bezeichnet - ist einer der wesentlichen leistungsbegrenzenden Faktoren heutiger zellulärer Mobilfunknetze. Seine sorgfältige statistische Beschreibung ist demnach ausschlaggebend für die Systemanalyse und den Systementwurf. Von einem Systemmodell zur Interferenz-Analyse wird verlangt, grundlegende Auswirkungen von Veränderungen im Netzwerk, wie etwa die Stationierung der Basisstationen und die Charakteristiken der Signalausbreitung, abzubilden. Darüber hinaus sollte es unkompliziert und flexibel sein, um einen ersten Einblick auf Entwurfsgrundlagen zu gewähren und den raschen Austausch von Ideen zu ermöglichen. Die Modellierung von Interferenz stellt bereits seit der Errichtung traditioneller makro-zellulärer Netzwerke eine Herausforderung dar. Mit dem jüngsten Aufkommen heterogener Netzwerk-Topologien wird die Situation zusätzlich erschwert, da viele der wohletablierten Aspekte herkömmlicher Methoden in Frage gestellt werden.

In dieser Dissertation werden benutzerzentrische System-Modelle vorgestellt, welche es ermöglichen, Szenarien mit asymmetrischer Interferenz-Einwirkung zu untersuchen. Der erste Ansatz vereinfacht die Interferenz-Analyse in einem hexagonalen Rastermodell, indem die Leistung der störenden Basisstationen gleichmäßig entlang eines Kreises verteilt wird. Die aggregierte Interferenz wird durch eine einzige gammaverteilte Zufallsvariable modelliert. Ihre Form- und Skalierungsparameter werden dabei über die Geometrie des Netzwerks sowie über den Schwund ermittelt. Das zweite Modell erweitert das zirkuläre Konzept um ungleichmäßige Leistungsprofile entlang der Kreise. Es ermöglicht die Abbildung beträchtlich großer Verteilungen von heterogenen Störern außerhalb der Zelle auf eine wohldefinierte, zirkuläre Anordnung von Netzelementen. Dabei reduziert es die Komplexität erheblich, während die ursprüngliche Interferenz-Verteilung erhalten bleibt. Das Modell wird durch eine neue finite Summen-Repräsentation für Gamma Zufallsvariablen mit ganzzahligem Formparameter ergänzt, welche es erlaubt, unter den Basisstationen Kandidaten für benutzerzentrische Basisstations-Kollaborationsschemen zu identifizieren und die dazu entsprechenden Durchsatzraten vorauszusagen. Der dritte Ansatz bedient sich stochastischer Geometrie um zweischichtige, heterogene Netzwerke unter Bezugnahme auf die Topologie einer urbanen Umgebung zu modellieren. Er löst die Problematik asymmetrischer Interferenz durch eine Näherung mittels eines virtuellen Gebäudes und stellt darüber hinaus ein neues Signalausbreitungs-Modell vor, das einen direkten Zusammenhang zu den Charakteristiken der Topologie wie etwa Gebäudedichte und Gebäudegröße herstellt, welche mühelos aus realen Daten extrahiert werden können.

Im letzten Teil der Dissertation wird die Verwendbarkeit der vorgestellten Modelle mittels Simulationen mit dem Vienna LTE-Advanced Dowlink System Level Simulator validiert. Dazu werden die analytischen Modelle mit Resultaten von LTE-Advanced Link Level Simulationen kalibriert. Dieser Abschnitt ergänzt zudem die bis hierhin auf den Benutzer fixierten Betrachtungen mit systemumfassenden Leistungsevaluierungen und behandelt insbesondere den Einfluss räumlicher Benutzer-Anhäufungen sowie der Dichte und Abschottung von Femtozellen. Der Fokus wird dabei besonders auf eine systematische und reproduzierbare Simulations-Methodik sowie geeigneten Leistungsmetriken gelegt, da konventionelle Leistungszahlen dazu neigen, Ungleichgewichte zwischen Benutzern zu verbergen.

Contents

1. **Motivation**   1
   - 1.1. Outline 4
   - 1.2. Notation 6

2. **Modeling Interference in Heterogeneous Cellular Networks**   7
   - 2.1. Stochastic Geometry 8
     - 2.1.1. Point Process Theory 8
     - 2.1.2. Poisson Point Process 10
     - 2.1.3. Probablity Generating Functional 10
     - 2.1.4. Techniques for Network Performance Analysis 11
   - 2.2. Interference Approximation by Known Probability Distributions 12
     - 2.2.1. Candidate Distributions 13
     - 2.2.2. Distance Metrics 14
     - 2.2.3. The Gamma Distribution 14

3. **Modeling Regular Aggregate Interference by Symmetric Structures**   17
   - 3.1. Hexagonal Reference Model 18
   - 3.2. Circular Interference Model 19
     - 3.2.1. Proposed Model 20
     - 3.2.2. The Dual Model 20
   - 3.3. Statistics of Aggregate Interference 21
     - 3.3.1. Interference Statistics at the Center 21
     - 3.3.2. Interference Statistics outside the Center 22
   - 3.4. Numerical Results and Discussion 23
     - 3.4.1. Validation of Expected Aggregate Interference 24
     - 3.4.2. Validation of Gamma Approximation 25
   - 3.5. Application in Heterogeneous Networks 28
   - 3.6. Summary 31

4. **Modeling Asymmetric Aggregate Interference by Symmetric Structures**   35
   - 4.1. Circular Interference Model 36

4.2. Distribution of the Sum of Gamma Random Variables 40
       4.2.1. Proposed Finite Sum Representation 41
       4.2.2. Application in Circular Interference Model 42
  4.3. Mapping Scheme for Stochastic Network Deployments 43
       4.3.1. Mapping Procedure 43
       4.3.2. Performance Evaluation for Homogeneous Base Station Deployments . 45
       4.3.3. Performance Evaluation of Heterogeneous Base Station Deployments . 48
       4.3.4. Power Profiles of Poisson Point Process (PPP) Snapshots 49
  4.4. Interference and Rate at Eccentric User Locations 50
       4.4.1. Generic Circularly Symmetric Scenario 50
       4.4.2. Components of Asymmetric Interference 51
       4.4.3. Transmitter Collaboration Schemes 53
  4.5. Summary 57

5. **Analysis of Urban Two-Tier Heterogeneous Cellular Networks**   **59**
  5.1. Preliminaries 60
       5.1.1. Random Shape Theory 60
       5.1.2. Indoor Coverage Ratio 61
  5.2. System Model 62
       5.2.1. Topology Model for Urban Environments 62
       5.2.2. Network Deployment 62
       5.2.3. User Association 63
       5.2.4. Virtual Building Approximation 64
       5.2.5. Signal Propagation 65
  5.3. Performance Analysis 66
       5.3.1. Typical Building with Small Cell Base Station (BS) 66
       5.3.2. Typical Building without Small Cell BS 67
       5.3.3. Typical Indoor User 68
  5.4. Numerical Evaluation 69
  5.5. LOS- and NLOS Macro Base Stations 71
       5.5.1. Distance Distributions of Associated Macro Base Stations 72
       5.5.2. SINR and Coverage Analysis 74
       5.5.3. Numerical Evaluation 76
  5.6. Summary 79

6. **LTE-A System Level Simulations**   **81**
  6.1. Vienna LTE-A Downlink System Level Simulator 81
  6.2. Homogeneous Macro Cellular Network 85
       6.2.1. System Model 85

		6.2.2. Validation of Gamma Approximation	86
		6.2.3. Validation of Asymmetric Interference Impact	89
	6.3.	Two-tier Heterogeneous Cellular Network	93
		6.3.1. System Model	93
		6.3.2. Urban Two-tier Heterogeneous Cellular Network	96
		6.3.3. User Hot Spot Scenarios	99
		6.3.4. Sensitivity on Femtocell Deployment Density and -Isolation	101
	6.4.	Summary	106

7. Conclusions 109
7.1. Summary of Contributions 109
7.2. Open Issues and Outlook 111
7.3. Conclusion 112

A. List of Abbreviations 113

B. Limitation of the Gaussian Approximation for the Aggregate Interference 115

C. Proof of Theorem 4.1 117

D. *Mathematica®* Code for Theorem 4.1 121

E. Proof of Theorem 5.1 123

F. Proof of Theorem 5.3 125

Bibliography 127

Chapter 1.

Motivation

Numerous studies predict an exponential growth of mobile data traffic for the next decade [1–3]. They follow an observation from Martin Cooper, who stated in 2001 that *"wireless network capacity doubles every 30 months"* [4]. Around the year 2010, this enormous growth rate triggered a paradigm shift in the design of wireless cellular networks. It became evident that the existing homogeneous macro-cell topologies will not be capable of sustaining these ever increasing demands. The time had come to end the *era of coverage* and to herald the *era of capacity*. Cooper's law assessed that wireless capacity has improved by a factor of one million during the last 45 years. The major part, a tremendous factor of 1600, has been achieved by efficient spatial reuse of spectrum [5, 6], as illustrated in Figure 1.1. However, adding further macro-cells was not a viable option due to cost and lack of available sites [7].

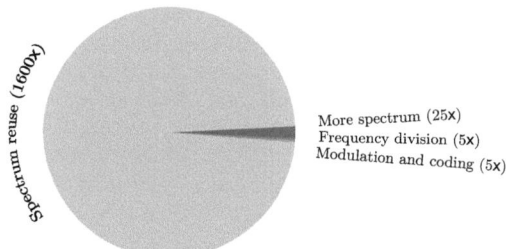

Figure 1.1.: Breakdown of one million times improvement (1600 × 25 × 5 × 5) in wireless network capacity during the last 45 years, as observed by Martin Cooper [4].

Chapter 1. Motivation

In response, BS deployments became increasingly heterogeneous, encompassing a wide range of transmit powers, carrier frequencies, backhaul connection types and communication protocols [8–10]. Both 3GPP LTE- and IEEE WiMAX standard added the notion "*small cell*" as an umbrella term for low-cost low-power radio access nodes that operate in a range of several meters to several hundred meters [11–15]. A network, which is constituted by traditional macro-BSs and embedded small cells, as shown in Figure 1.2, is called a *multi-tier-* or *heterogeneous* cellular network[1], respectively.

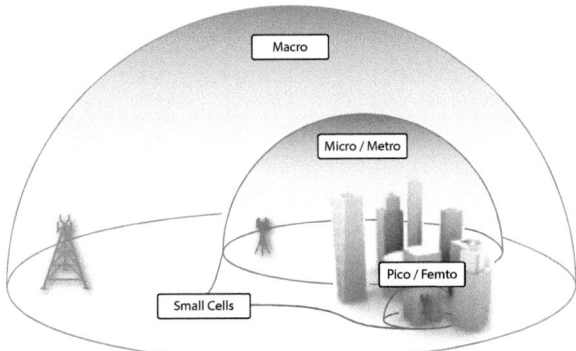

Figure 1.2.: Heterogeneous cellular network in urban environment. "*Small cell*" serves as an umbrella term for micro-, metro-, pico- and femto cells, respectively.

In such multi-tier topologies, *aggregate interference*[2], i.e., the cumulative impact of all co-channel interferers, is one of the main performance limiting factors [6, 19–25]. Hence, its careful statistical characterization is crucial for system design and analysis. Interference modeling has been a challenging problem ever since the emergence of traditional single-tier cellular networks. The main target is to capture key dependencies of the interference as a function of relatively few parameters [26]. A widely used approach is the application of Gaussian random processes [27–31]. This model is justified when accumulating a large number of interference contributions without a dominant term[3] such that the Central Limit Theorem (CLT) applies [32–36]. In many cases, however, the Probability Density Functions (PDFs) will exhibit heavier tails than those

---

[1] In this thesis, the term "*tier*" either refers to a ring of BSs in a hexagonal-grid setup or the specific part of a network, which is associated with a certain class of BSs, such as macro-BSs and small cells, respectively. The particular meaning becomes apparent from the context.

[2] Also referred to as *(aggregate) network interference* in literature [16–18].

[3] The scenarios as presented in this thesis violate the conditions for the validity of the Gaussian approximation, as shown in Appendix B.

predicted by the Gaussian approach [19, 37–42].

The two main interference shaping factors are the location of the BSs and the path loss law [16, 19, 26, 43–48]. Thus, interference is determined to first order by the network geometry or more precisely, the interference geometry, which condenses the underlying BS distribution and the channel access scheme [49, 50]. The path loss law models the distance dependent signal attenuation.

BS locations are typically abstracted to some baseline model, since the main aspects of a system should hold across a wide range of deployment scenarios. For more than three decades the hexagonal grid has been extensively used by both academia and industry [51–55]. It is flexible and commonly accepted as a reasonably useful model to represent well-planned homogeneous BS deployments, which made it withstand the test of time [8, 56]. On the other hand, it is generally not tractable, thus requiring overly simplistic assumptions or extensive simulations [21, 57–60]. Such simulations are not only tedious to implement and run, but also often lack openly accessible source code for reproducibility and a mathematical backup. As a result, they rarely inspire smart new algorithms and designs.

In the context of heterogeneous networks, small cell locations are typically beyond the scope of network planning and, thus, more of a random nature [7, 8, 10, 21, 61–65]. Without prior information, the best statistical model is a uniform distribution, which corresponds to complete spatial randomness [66]. In that case, network properties can conveniently be captured by a Point Process (PP) and enable to leverage techniques from *stochastic geometry*. This powerful mathematical framework recently gained momentum as the only available tool that provides - due to its random nature perhaps counterintuitively - a rigorous approach to modeling, analysis and design of multi-tier cellular networks [16, 19, 22, 23, 26, 43, 47, 49, 67–73]. However, when closed-form expressions are desired, it imposes its own particular limitations [14, 47, 74], typically including spatial stationarity and isotropy of the scenario. Hence, notions such as *cell-center* and *cell-edge* are, in general, not accessible.

Spatial randomness represents the philosophical opposite of the regular grid. As a result, these two extremes provide upper- and lower performance bounds for any conceivable heterogeneous network deployment [7]. The gap is greatest, when the distance between desired BS and receiver resembles the typical inter-BS distance [75]. The main objective of this thesis is to reduce the difference between the performance boundaries at arbitrary receiver locations. Firstly, it aims to represent unplanned topologies by more regular structures. The goal is to accurately capture characteristics of a heterogeneous network deployment, such as the presence of dominant interferers, while preventing the receiver from being located arbitrarily close to an interfering node. The second part of the thesis focuses on the fact that an urban environment naturally imposes certain repulsion between the transmitters. It approaches the lower boundary,

Chapter 1. Motivation

corresponding to spatial randomness, by embedding stochastic models into more realistic environment topologies. The third pillar of the work addresses methods towards systematic- and reproducible system level simulations of heterogeneous networks. It aims to raise awareness on performance metrics, which are undergoing a paradigmatic change in multi-tier topologies.

1.1. Outline

This thesis is structured as follows. Chapter 2 provides an introduction on modeling interference in heterogeneous cellular networks. The second part of the thesis, spanning chapters 3 to 5, presents theoretical models for investigating asymmetric aggregate interference at a certain user location. It addresses both homogeneous regular- as well as heterogeneous stochastic BS deployments. Chapter 6 complements the work by comprehensive system level simulations, both validating and extending the results from theory. The remainder of this section provides a short abstract of each chapter and exclusively refers to literature, which was published by the author.

Chapter 2

This chapter provides preliminaries so as to make the concepts of this thesis more accessible. It introduces state of the art models for the analysis of heterogeneous cellular networks. Particular light is shed on stochastic geometry, including key results and important assumptions. The second part of the chapter outlines methods to approximate aggregate interference statistics by well-known PDFs.

Chapter 3

Before the advent of random spatial models, the hexagonal grid was favored as a reasonably useful model for regularly-arranged homogeneous cellular network topologies. However, its geometry renders the analysis of aggregate interference statistics *at eccentric user locations* difficult in general. Based on my contributions in [76], in this chapter, it is proposed to approximate the aggregate interference by a Gamma Random Variable (RV) and to exploit a circular interference model with uniform power profile for determining the corresponding shape and scale parameters. Its applicability for representing the well-planned part of a heterogeneous cellular network is demonstrated.

Chapter 4

Referring to my work in [77], in this chapter, the circular model is extended to multiple circles with non-uniform power profiles, allowing to map *arbitrary network topologies* on the circular model. In this approach, asymmetric interference can be caused by both an eccentric user location and an unequal impact of the interferers. A heuristic mapping scheme for heterogeneous cellular networks is demonstrated. In order to make the analysis more convenient, a new theorem for calculating the distribution of a sum of Gamma RVs with integer shape parameter is introduced. It enables to detect candidate base stations for user-centric base station collaboration schemes and allows to predict the corresponding Signal-to-Interference plus Noise Ratio (SINR)- and rate performance.

Chapter 5

Yet, numerous studies have analyzed the characteristics of heterogeneous cellular networks, most of them based on stochastic geometry. The signal propagation environment is typically modeled as a flat plane without any obstacles. However, heterogeneous cellular networks are most qualified to operate in dense urban environments. Based on my contributions in [78], in this chapter, the deficiency is overcome by a tractable model for urban environment topologies. Asymmetric aggregate interference is addressed by a virtual building approximation.

Chapter 6

Chapters 3 to 5 present analytical models to evaluate performance from a *user-centric* point of view. The first goal of this chapter is to validate the obtained results with the Vienna LTE-A system level simulator. Then, it is aimed at the big picture by investigating *system-wide* performance. Based on my work in [79, 80], the impact of user clustering as well as small cell density and -isolation are addressed. A particular focus is placed upon a systematic- and reproducible simulation methodology. Further emphasis is put on appropriate performance metrics, since conventional figures of merit, as used in single-tier networks, tend to conceal severe performance imbalances among users. Related work on the Vienna LTE-A simulator is presented in [81–85].

1.2. Notation

The following notation is used throughout this thesis.

Table 1.1.: Mathematical notation.

Symbol	Annotation
$f_X(\cdot)$	Probability density function of X
$F_X(\cdot)$	Cumulative distribution function of X
$\mathbb{E}[X]$	Expected value of X
$\text{Var}[X]$	Variance of X
$\Gamma[k, \theta]$	Gamma distribution with shape k and scale θ
$\mathbf{v} \in \mathbb{R}^{d \times 1}$	Real-valued column vector of length d
$[\mathbf{v}]_l$	l-th element of vector $\mathbf{v}$
Φ	Point process on $\mathbb{R}^d$
$\mathcal{B}(x, r)$	Ball with center $x \in \mathbb{R}^2$ and radius $r > 0$

The subsequent table summarizes commonly employed parameters.

Table 1.2.: Frequently used parameters.

Symbol	Annotation
(r, ϕ)	User location in polar coordinates
(R, Ψ)	Transmitter-site location in polar coordinates
P_M	Macro base station transmit power in $[\text{W}]$
μ_M	Macro base station density in $[m^{-2}]$
P_S	Small cell transmit power in $[\text{W}]$
μ_S	Small cell density in $[m^{-2}]$
η	Small cell occupation probability/-ratio, $0 \leq \eta \leq 1$
$\ell(\cdot)$	Distance-dependent path loss law, $0 \leq \ell(\cdot) \leq 1$
R_I	Radius of indoor area or building
L_W	Wall penetration loss, $0 \leq L_\text{W} \leq 1$
S, I	Aggregate signal- and interference powers in $[\text{W}]$
γ	Signal-to-Interference Ratio, $\gamma = S/I$
τ	Normalized rate (spectral efficiency) in $[\text{bit/s/Hz}]$

Chapter 2.

Modeling Interference in Heterogeneous Cellular Networks

Aggregate interference[1] is one of the main performance limiting factors in today's wireless cellular networks, making its accurate statistical characterization imperative for network design and analysis. The two dominant interference-shaping factors are the spatial distribution of concurrently transmitting Base Stations (BSs) and the path loss, which encompasses signal attenuation by distance and fading [16, 19, 43, 46, 47, 86].

Although abstraction models such as the Wyner model and the hexagonal grid have been reported two- or even five decades ago [87, 88], mathematically tractable interference statistics are still the exception rather than the rule. Moreover, the recent emergence of heterogeneous network topologies complicates matters by fundamentally challenging various time-honored aspects of traditional network modeling [10].

In current literature, BS deployment models mainly follow the trend away from being fully deterministic towards complete spatial randomness [8]. However, even with the new models, only particular combinations of spatial node distributions, path loss models and receiver locations yield known expression for the Probability Density Function (PDF) of the aggregate interference [14], and hence allow to predict the statistics of further performance metrics such as Signal-to-Interference plus Noise Ratio (SINR), outage and rate. For example, a finite number of interferers together with certain fading distributions, such as Rayleigh, lognormal or Gamma, allows to exploit a vast amount of literature on the sum of Random Variables (RVs) [89–109]. Otherwise, tractable interference statistics have mainly been reported in the field of stochastic geometry.

---

[1]Throughout this thesis, the term *aggregate interference* is used to denote the cumulative impact of all co-channel interferers in order to clearly distinguish it from the *interference* of a single source.

In the general case, the PDF is unknown and aggregate interference is typically characterized by either the Laplace Transformation (LT), the Characteristic Function (CF) or the Moment Generating Function (MGF), respectively [14]. In this thesis, the LT is considered most relevant due to its suitability for RVs with non-negative support and its moment generating properties. Let I denote a RV with PDF $f_I(x)$, representing the aggregate interference. Then, its LT is given as

$$\mathcal{L}_I(s) = \mathbb{E}\left[e^{-sI}\right] = \int_0^\infty f_I(x) e^{-sx} dx. \tag{2.1}$$

The n-th moment of I is determined by

$$\mathbb{E}\left[I^n\right] = (-1)^n \left.\mathcal{L}_I^{(n)}(s)\right|_{s=0}, \tag{2.2}$$

where $\mathcal{L}_I^{(n)}(s)$ refers to the n-th derivative of $\mathcal{L}_I(s)$.

The next section introduces fundamental principles of stochastic geometry, which provides means to systematically evaluate the LT.

2.1. Stochastic Geometry

Stochastic geometry analysis is based on the concept of abstracting BS deployments to Poisson Processes (PPs). It encompasses a framework that outputs spatial averages for the figures of merit over a large number of network realizations [14]. In wireless communication engineering, stochastic geometry has been considered as early as 1997[2] [22, 23, 67]. However, key metrics such as coverage and rate have not been determined at this time.

2.1.1. Point Process Theory

Let N be the set of all sequences of points in $\mathbb{R}^d$, such that any sequence $\phi \in$ N is (i) finite, i.e, any bounded subset $B \subset \mathbb{R}^d$ contains only a finite number of points, and (ii) simple, i.e., $x \neq y$ for any $x, y \in \phi$. Define $\phi(B)$, with $\phi \in$ N and $B \subset \mathbb{R}^d$ as the number of points of ϕ in B. Further, let $\mathcal{N}$ be the smallest sigma algebra, such that the mappings $\phi \to \phi(B)$ are measurable for all Borel subsets $B \subset \mathbb{R}^d$. Finally, consider a probability space $(\Omega, \mathcal{F}, \mathbb{P})$, where Ω denotes the sample space, $\mathcal{F}$ is the set of events and $\mathbb{P}$ corresponds to the mapping, which assigns probabilities to the events.

---

[2]In a planar network of nodes, which are distributed according to an arbitrary PP, interference can be modeled by a generalized shot noise process [110, 111]. Hence, the roots of stochastic geometry date back even further, referring to shot noise studies of Campbell in 1909 [112, 113], and Shottky in 1918 [114].

2.1. Stochastic Geometry

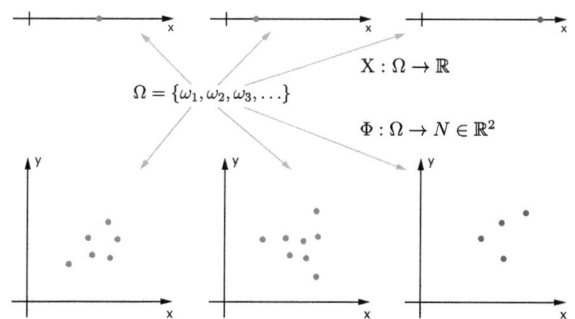

Figure 2.1.: RV on $\mathbb{R}$ (upper part of figure) vs. PP on $\mathbb{R}^2$ (lower part of figure).

Definition 2.1. *A point process Φ on $\mathbb{R}^d$ is a measurable mapping from a probability space $(\Omega, \mathcal{F}, \mathbb{P})$ to $(\mathsf{N}, \mathcal{N})$, i.e.,*

$$\Phi : \Omega \to \mathsf{N}. \tag{2.3}$$

The definition is referred from [19, Appendix A] and informally describes a PP as a random variable, which takes values from the set of simple and finite sequences N on $\mathbb{R}^d$. If the PP is ergodic, the spatial average, i.e, across points, and the ensemble average, i.e, across realizations are equal [71]. Figure 2.1 illustrates the difference between a RV on $\mathbb{R}$ and a PP on $\mathbb{R}^2$.

Definition 2.2. *The intensity measure Λ of a point process Φ is defined as*

$$\Lambda(B) = \mathbb{E}[\Phi(B)], \quad \forall B \subset \mathbb{R}^d. \tag{2.4}$$

According to [19, Appendix A], it equals the average number of points in $B \subset \mathbb{R}^d$.

Definition 2.3. *A point process $\Phi = \{x_n\}$ is called stationary if $\Phi_x = \{x_n + x\}$ has the same distribution as Φ for all $x \in \mathbb{R}^d$, i.e.,*

$$\mathbb{P}[\Phi \in Y] = \mathbb{P}[\Phi_x \in Y], \quad \forall Y \in \mathcal{N}, \tag{2.5}$$

and it is called isotropic if

$$\mathbb{P}[\Phi \in Y] = \mathbb{P}[\Phi_x \in rY], \tag{2.6}$$

where r is a rotation in $\mathbb{R}^d$.

2.1.2. Poisson Point Process

The most well-studied and most widely used PP is the Poisson Point Process (PPP). Its importance mainly results from its *independence property* [14, 43, 68].

Definition 2.4. *A PP* $\Phi = \{x_i;\ i = 1, 2, 3, \ldots\} \subset \mathbb{R}^d$ *is a PPP iff the number of points within any compact set* $B \subset \mathbb{R}^d$ *is a Poisson RV, and the number of points in disjoint sets are independent.*

If $\Lambda(B) = \lambda|B|$, the PPP is called *homogeneous* with intensity λ. The independence property relates to the fact that there is no dependence between point locations. While this appears to be a reasonable assumption for the unplanned deployment of small cells, it is disputable for macro-BS locations, where a regular-grid or a repulsive PP[3] might better reflect the basic planning procedure [7, 61, 69, 115, 116].

In this regard, the PPP, shows a broad analogy to wireless channel modeling by Rayleigh fading [10]. Although it is commonly known that Rayleigh is not particularly accurate, it captures essential channel variation effects and first-order insights on many wireless design fundamentals, including diversity and multi-antenna transmission. Its simplicity and tractability enabled wireless communication engineers to rapidly exchange new ideas. The PPP model has broadly comparable characteristics in the context of heterogeneous network topologies with a high degree of spatial randomness. It enables to evaluate the Probability Generating Functional (PGFL), which completely characterizes a simple PP and, under certain restrictions, eventually leads to closed form expressions for the PDF of the aggregate interference [19].

An interesting property of PPPs is *thinning*. It describes the procedure of independently selecting a point of the process with probability v and discarding it with probability $1-v$. This results in two independent PPPs with intensity measures $v\Lambda$ and $(1-v)\Lambda$, where $(v\Lambda)(B) = \int_B v(x)\Lambda(dx)$ and $((1-v)\Lambda)(B) = \int_B (1-v(x))\Lambda(dx)$, respectively.

2.1.3. Probablity Generating Functional

Definition 2.5. *The PGFL of a point process* Φ *is defined as*

$$\mathcal{G}[g] = \mathbb{E} \prod_{\mathsf{x}\in\Phi} g(\mathsf{x}),$$

where $g(x) : \mathbb{R}^d \to [0, \infty)$ *is measurable*[4].

---

[3] A repulsive PP is a variation of the PPP, which enforces a minimum distance between node locations.
[4] The symbol 'x' is used to distinguish the actual points of the PP from arbitrary points x in $\mathbb{R}^d$.

2.1. Stochastic Geometry

For a PPP, the PGFL equals

$$\mathcal{G}[g] = \exp\left(-\int_{\mathbb{R}^d}(1-g(x))\Lambda(dx)\right).$$

It proves particularly useful to evaluate the LT of the sum $\sum_{x\in\Phi} f(x)$:

$$\mathbb{E}\left[\exp(-s\sum_{x\in\Phi} f(x))\right] = \mathbb{E}\left[\prod_{x\in\Phi} \exp(-sf(x))\right]$$

$$= \mathcal{G}[\exp(-sf(\cdot))], \qquad (2.7)$$

which typically appears in the analysis of aggregate interference, where $f(\cdot)$ represents the contribution from a single node, i.e., $I = \sum_{x\in\Phi} f(x)$.

2.1.4. Techniques for Network Performance Analysis

When the nodes of a homogeneous BS deployment are uniformly scattered over the infinite plane and the fading is represented by independent and identically distributed (i.i.d.) non-negative RVs, the PDF of the aggregate interference yields a *skewed stable* distribution [16, 19, 117]. Yet, this is the only available case in literature, which leads to known statistics. Still, except for a Lévy distribution, which is obtained by assuming a path loss exponent of 4, it does not result in any closed-form expressions for the PDF [14].

To overcome these obstacles, various methods have been reported in literature that commonly exploit the LT of the aggregate interference, $\mathcal{L}_I$, as a basis for calculating further performance metrics [14]:

1. Resort to Rayleigh fading on the desired link. Then, according to [14], the exact distribution of the SINR γ is obtained as

$$F_\gamma(\delta) = \mathbb{P}[\gamma > \delta]$$

$$= 1 - \exp(-sW)\mathcal{L}_I(s)|_{s=c\delta}, \qquad (2.8)$$

where W denotes the noise power. Given the exact Cumulative Distribution Function (CDF) of the SINR, the coverage probability $P_c(\delta) = \mathbb{P}[\gamma < \delta]$ is readily obtained by the Complementary Cumulative Distribution Function (CCDF). According to [21, 73], the normalized ergodic rate, $\tau = \mathbb{E}[\log_2(1+\gamma)]$, can be expressed as

$$\tau = \frac{1}{\log 2} \int_{\delta > 0} \frac{P_c(\delta)}{\delta + 1} d\delta. \tag{2.9}$$

This approach is applied in numerous studies [21, 58, 74, 118–124]. Despite its convenient tractability, neglecting the shadow fading misses some important dynamics of a dense urban environment, as presented in Chapter 5.

2. Resort to dominant interferers by region bounds or nearest n interferers. This yields a lower bound on the outage probability, as employed, e.g., in [72, 125].

3. Resort to Plancherel-Parseval theorem. Let $f_1(t)$ and $f_2(t)$ be square integrable complex functions. Then, from [126],

$$\int_{\mathbb{R}} f_1(t) f_2^*(t) dt = \int_{\mathbb{R}} \mathcal{F}_1(\omega) \mathcal{F}_2^*(\omega) d\omega \tag{2.10}$$

This method omits to calculate the inverse Laplace transform and is applicable to general fading environments, as reported, e.g., in [26, 49, 74]. However, it yields intricate integrals and hence often falls short of simple closed form analysis.

4. Calculate the aggregate interference PDF by directly inverting the LT, CF or MGF. With few exceptions, such as reported in [16, 117, 127, 128], the inversion has to be carried out numerically [39, 129] and obstructs the path to closed-form expressions.

5. Resort to the approximation of the aggregate interference PDF by a known distribution and determine its parameters by LT, CF or MGF. This method is not only employed to make non-isotropic interference and non-Poissonian PPs, such as hardcore- and cluster processes analytically accessible [17, 130], but is also heavily used outside the stochastic domain [46, 131–133]. Therefore, it is discussed in greater detail in the next section.

2.2. Interference Approximation by Known Probability Distributions

It is well-studied that the Central Limit Theorem (CLT) and the corresponding Gaussian model provide a very poor approximation for modeling aggregate interference statistics in large wireless cellular networks [16, 130, 134]. Its convergence can be measured by the Berry-Esseen inequality [135] and is typically thwarted by a few strong interferers, as shown in Appendix B. The resulting PDF exhibits a heavier tail than what is anticipated by the Gaussian model [16].

2.2. Interference Approximation by Known Probability Distributions

Resorting to the approximation of the aggregate interference distribution by a known parametric distribution imposes two challenges: (i) the choice of the distribution itself, and (ii) the parametrization of the selected distribution. Although there is *no known criterion* for choosing the optimal PDF, its tractability for further performance metrics as well as the characteristics of the spatial model, path loss law and fading statistics advertise certain candidate distributions.

2.2.1. Candidate Distributions

The characteristics of the interference distribution, such as skewness and kurtosis are, to a certain extent, dependent on the receiver's position within the cell. A major motivation for the investigations in this thesis stems from the fact that spatial averaging tends to obscure these location-specific uncertainties, and is thus argued to limit the insights by stochastic geometry modeling [129, 136].

As indicated in Section 2.1, the only yet existing fully-tractable stochastic geometry setup results in a skewed stable distribution, in particular a Lévy distribution [137]. The Lévy distribution is a special case of the inverse-Gamma distribution, which belongs to the class of generalized inverse-Gaussian distributions. Another class of continuous probability distributions that allows for positive skewness and non-negative support are normal variance-mean mixtures, in particular the normal inverse-Gaussian distribution. The main penalty of such generalized distributions is the need to determine up to four parameters, which typically exhibit non-linear mappings when applying moment- or cumulant matching [46]. Hence, it is beneficial to resort to special cases with only two parameters. Inverse-Gamma- , inverse-Gaussian-, log-normal- and Gamma have frequently been reported to provide an accurate abstraction of the aggregate interference statistics, e.g., in [46, 130, 133], [46, 130], [138–142] and [17, 46, 143–145], respectively.

Ideally, the distribution parameters are directly obtained as a function of the scenario parameters, such as BS density, path loss exponent and variance of the fading. This thesis only embraces fading distributions without a heavy tail[5] and scenarios that claim a certain minimum distance between transmitter and receiver. It follows immediately that the aggregate interference distribution does not exhibit a heavy tail either, as shown, e.g., in [19], and that all moments exist and are finite. This allows to exploit moment- and cumulant matching methods, e.g., [17, 46, 133, 142–150] and [132, 140, 151–154], as an intermediate step to determine the mapping between scenario- and distribution parameters. Otherwise, numerical distribution fitting has to be applied [155], which is not within the scope of this work.

---

[5]A distribution is denoted as *heavy tailed*, if its tail is not exponentially bounded , i.e., $\lim_{x \to \infty} e^{ax} \mathbb{P}[X > x] = \infty, \ \forall a > 0$.

2.2.2. Distance Metrics

In order to quantify the goodness of fit between the actual interference distribution and its approximation, this work employs the *Kolmogorov-Smirnov (KS) test*. Let $F_I(x)$ and $F_0(x)$ denote two CDFs. Then, the KS test statistic is given as

$$D_{\text{KS}} = \sup_x |F_I(x) - F_0(x)|. \tag{2.11}$$

Alternatively, the *Kullback-Leibler divergence* is often applied as a distance measure [156]. For two distributions P and Q of a continuous RV, it is defined by

$$D_{\text{KL}} = \int_{-\infty}^{\infty} f_p(x) \log \frac{f_p(x)}{f_q(x)} dx, \tag{2.12}$$

with $f_p(x)$ and $f_q(x)$ referring to the PDFs of P and Q, respectively. However, the Kullback-Leibler divergence does not satisfy symmetry and triangle inequality, thus amounting to a *pre-metric*.

2.2.3. The Gamma Distribution

In this work, particular focus is placed upon the Gamma distribution due to its wide range of useful properties for wireless communication engineering, some of which are outlined in this section.

The PDF of a Gamma distributed RV X with *shape parameter* k and *scale parameter* θ, i.e., $G \sim \Gamma[k, \theta]$, is defined as

$$f_G(x) = \frac{1}{\theta^k \Gamma(k)} x^{k-1} e^{-x/\theta}. \tag{2.13}$$

Its mean and variance are given by $\mathbb{E}[G] = k\theta$ and $\text{Var}[G] = k\theta^2$.

The Gamma distribution exhibits the *scaling property*, i.e., if $G \sim \Gamma[k, \theta]$, then $aG \sim [k, a\theta]$, $\forall a > 0$, and the *summation property*, i.e., if $G_i \sim \Gamma[k_i, \theta]$ with $i = 1, 2, \ldots, N$, then $\sum_{i=1}^{N} G_i \sim \Gamma[\sum_{i=1}^{N} k_i, \theta]$.

Consider an arbitrary distribution with mean ν and variance σ^2. Then, the distribution $\Gamma[k, \theta]$ with the same first- and second order moments has the parameters

$$k = \frac{\nu^2}{\sigma^2}, \quad \theta = \frac{\sigma^2}{\nu}. \tag{2.14}$$

2.2. Interference Approximation by Known Probability Distributions

These simple moment-matching identities can be exploited for accurately approximating fading distributions [146], such as generalized-K [148, 149] and log-normal [146–149], as well as aggregate interference statistics [17, 144, 145].

Additionally, the Gamma distribution covers the power fading distribution of various single- and multi-antenna schemes under the Rayleigh fading assumption. Conventional Single-Input-Single-Output (SISO) yields an exponential distribution $\text{Exp}[1/\theta]$, which is equivalent to $\Gamma[1,\theta]$. The power fading of Maximum Ratio Transmission (MRT) with N_{Tx} transmit antennas and one receive antenna can be modeled by $\Gamma[N_{\text{Tx}}, \theta]$, Maximum Ratio Combining (MRC) with one transmit antenna and N_{Rx} receive antennas is characterized by $\Gamma[N_{\text{Rx}}, \theta]$. Furthermore, MRC is often studied in the presence of Nakagami-m fading. Let $Y \sim \text{Nakagami}[m, \Omega]$ and $G = Y^2$. Then, $G \sim \Gamma[m, \Omega/m]$.

According to [157], the quotient $\gamma = S/I$ of two RVs $S \sim \Gamma[k_S, \theta_S]$ and $I \sim \Gamma[k_I, \theta_I]$ is distributed as

$$f_\gamma(x) = \frac{(\theta_I/\theta_S)^{k_S}}{B(k_S, k_I)} \left(1 + \frac{\theta_I}{\theta_S} x\right)^{-k_S - k_I} x^{k_S - 1}, \quad x > 0 \tag{2.15}$$

with $B(\cdot, \cdot)$ denoting the Beta function. Interpreting γ as a Signal-to-Interference Ratio (SIR) allows to determine the success probability $\mathbb{P}[\gamma > \delta]$ for a given threshold δ as

$$\mathbb{P}[\gamma > \delta] = \frac{\Gamma(k_S + k_I)}{\Gamma(k_S)} \left(\frac{\theta_S}{\delta \theta_I}\right)^{k_I} {}_2\bar{F}_1\left(k_I, k_S + k_I, 1 + k_I, -\frac{\theta_S}{\delta \theta_I}\right), \tag{2.16}$$

where ${}_2\bar{F}_1(\cdot, \cdot, \cdot, \cdot)$ is a regularized hypergeometric function [17].

These observations motivate the application of the Gamma RV as a sensible compromise between accuracy and tractability. Further properties of Gamma RVs will be discussed as needed throughout the course of this thesis.

Chapter 3.

Modeling Regular Aggregate Interference by Symmetric Structures

In this chapter, downlink co-channel interference statistics in wireless cellular networks with hexagonal grid layout are investigated. The main target is to facilitate the analysis at user locations outside the center of the cell of interest.

The proposal of a cellular structure for mobile networks dates back to 1947. Two Bell Labs engineers, Douglas H. Ring and W. Rae Young were the first to mention the idea in an internal memorandum [88]. Almost two decades later, in 1966, Richard H. Frenkiel and Philip T. Porter, shaped a *hexagonal cellular array of areas* to propose the first mobile phone system. Although never proposed as innovative research solution, the hexagon model gained high popularity within the research community and is still extensively utilized nowadays [158–163]. It serves either as the system model itself, or as a reference system for more involved simulation scenarios. On the other hand, its geometric structure renders closed-form analysis of *aggregate interference statistics* difficult [164]. Hence, simulation results often lack a mathematical back up.

Recently, closed-form results have been reported with system models based on stochastic geometry [17, 47, 71]. However, as detailed in Chapter 2, the stochastic approach is based on an ensemble of network realizations and is therefore not applicable when a fixed structure of the network is given. Since the *well-planned* deployment of macro-sites is not expected to vanish in the medium term, it is thus considered imperative to make interference analysis in the hexagonal grid model more convenient.

Current work on regular grid models has mainly focused on link-distance statistics [165, 166]. The authors also account for fading and provide closed-form approximations for the co-channel interference of a *single link*. However, convenient expressions for the moments and the distribution of *aggregate co-channel interference* are not available yet.

Chapter 3. Modeling Regular Aggregate Interference by Symmetric Structures

Based on my work in [76], the contributions of this chapter are:

- A *circular interference model* to facilitate interference analysis in cellular networks with regular grid layout is introduced. Particular focus is placed on the hexagonal grid due to its ubiquity in wireless communication engineering [158–163]. The key idea is to consider the power of the interfering BSs as being uniformly spread along the perimeter of the hexagon.

- It is proposed to model interference statistics in a hexagonal scenario by a single Gamma RV. Its shape- and scale parameters are determined in closed form by employing the *circular model*. The analysis yields key insights on the formative components of the interference distribution. A scenario with regularly arranged macro-sites and randomly distributed small cells demonstrates the model's expedient application in heterogeneous cellular networks.

The chapter forgoes hexagonal grid setups with more than one ring of interferers as well as further performance analysis, which is enabled by the Gamma approximation, as described in Section 2.1.4. Both are considered straightforward and of no particular relevance for this thesis.

3.1. Hexagonal Reference Model

The reference hexagonal setup is composed of a central cell and six interfering BSs, as shown in Figure 3.1. The interferers are equipped with omnidirectional antennas and are located at the edges of a hexagon with radius R (marked as '+' in Figure 3.1). All BSs are assumed to transmit with the same power. The signal from the i-th interfering BS with polar coordinates (R, Ψ_i) to a user with polar coordinates (r, ϕ) experiences macroscopic path loss and fading. It is assumed that $0 < r \leq R/2$, so as to assure that the user is associated with the central BS. The path loss is modeled by the exponential law

$$\ell\left(d_{r,\Delta_i}^{(M)}\right) = \min\left(b_\text{P}, \frac{1}{c_\text{P}}\left(d_{r,\Delta_i}^{(M)}\right)^{-\alpha}\right), \qquad (3.1)$$

where b_P denotes the intercept, c_P is a constant, α refers to the path loss exponent and

$$d_{r,\Delta_i}^{(M)} = \sqrt{R^2 + r^2 - 2Rr\cos(\Delta_i)}, \qquad (3.2)$$

with $\Delta_i = \phi - \Psi_i$ and $\Psi_i = 2\pi i/M$, $i = 1, \ldots, M$. In the remainder of this chapter, it is assumed that $d_{r,\Delta_i}^{(M)} > (b_\text{P} c_\text{P})^{-1/\alpha}$. Exemplifying from [167], a minimum coupling loss of 70 dB and free

3.2. Circular Interference Model

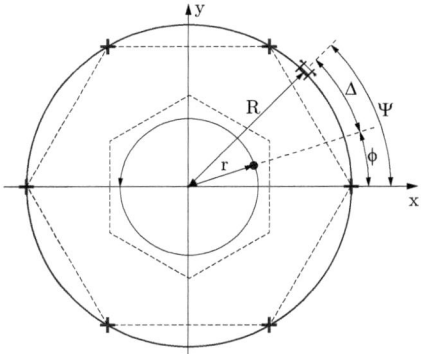

Figure 3.1.: System model. Center cell with user at (r,ϕ). Interfering BSs are located at (R,Ψ_i), where $\Psi_i = 2\pi\, i/M$, $i = 1,\ldots,M$.

space propagation at an LTE-A frequency of $f_c = 2.14\,\text{GHz}$ yield $b_\text{P} = 10^{-7}$, $c_\text{P} = 8.05 \cdot 10^{-3}$ and $(b_\text{P} c_\text{P})^{-1/\alpha} = 0.028\,\text{m}$, hence justifying this assumption.

In the hexagonal scenario, $M = 6$. The terms r and Δ_i denote the user's distance to the center and its angle-difference to the i-th interfering BS, respectively. Motivated by Section 2.2.3, fading is modeled by an i.i.d. Gamma RV $G_i \sim \Gamma[k_0, \theta_0]$, where k_0 and θ_0 refer to shape- and scale parameter, respectively.

3.2. Circular Interference Model

In a one-tier hexagonal grid scenario, as presented in Section 3.1, the experienced aggregate interference power at position (r,ϕ) can be expressed as

$$I_6(r,\phi) = \sum_{i=1}^{6} P_\text{M}\, G_i\, \ell\!\left(d^{(6)}_{r,\Delta_i}\right), \tag{3.3}$$

where P_M denotes the transmit power, G_i is the fading and $\ell(d^{(6)}_{r,\Delta_i})$ refers to the path loss at distance $d^{(6)}_{r,\Delta_i}$, as specified in (3.1) and (3.2), respectively. Each sum term can be regarded as a RV G_i, which is weighted by the factor $P_\text{M}\,\ell(d^{(6)}_{r,\Delta_i})$. Hence, the statistics of $I_6(r,\phi)$ outside the cell-center, i.e., $r > 0$, are accessible via a sum of differently weighted RVs. Since, in general, this does not lead to closed-form results, as detailed in Chapter 4, in this chapter a circular interference model is proposed in order to facilitate the statistical analysis.

3.2.1. Proposed Model

In the proposed circular interference model, the power of the six reference BSs is spread uniformly along a circle of radius R. This is achieved by evenly distributing the total transmit power $6\,P_\text{M}$ among M equally spaced BSs and considering the limiting case $M \to \infty$. By generalizing (3.3), this is expressed as

$$I_C(r) = \lim_{M\to\infty} \frac{6\,P_\text{M}}{M} \sum_{i=1}^{M} G_i\, \ell\left(d_{r,\Delta_i}^{(M)}\right) = \frac{6\,P_\text{M}}{2\pi} \mathbb{E}\left[G_i\right] \int_{-\pi}^{\pi} \ell(d_{r,\Delta})\, d\Delta, \qquad (3.4)$$

with $\ell(\cdot)$ from (3.1) and $d_{r,\Delta_i}^{(M)}$ from (3.2). The terms $d_{r,\Delta}$ and Δ denote distance and angle-difference between the user and an infinitesimal interfering circular segment, as illustrated in Figure 3.1.

Assuming a path loss exponent $\alpha = 2$, i.e., free space propagation, (3.4) can explicitly be evaluated as

$$I_C(r) = 6\,P_\text{M}\, \mathbb{E}\left[G_i\right] \frac{1}{c_\text{P}} \frac{1}{R^2 - r^2}. \qquad (3.5)$$

An intuitive interpretation of this result is provided in the next section by the model's pendant.

In the remainder of this chapter, $\alpha = 2$ is employed. It represents the worst case of low interference attenuation. However, previously- as well as all subsequently presented analysis can be carried out in closed-form for $\alpha = 2n$ with $n \in \mathbb{N}$. Values α other than these require the evaluation of elliptic integrals (see, e.g., [168]). Thus, a practical first order estimate for arbitrary values of α is achieved by evaluating the performance with $2n$ and $2(n+1)$, where $2n < \alpha < 2(n+1)$.

3.2.2. The Dual Model

Consider a user in a hexagonal scenario, which is moved along a circle of radius r from $-\pi$ to π, as indicated in Figure 3.1. The *average expected* interference along the circle can be calulated as

$$I'_C(r) = \frac{1}{2\pi} \int_{-\pi}^{\pi} \mathbb{E}\left[I_6(r,\phi')\right] d\phi' \qquad (3.6)$$

$$= \sum_{i=1}^{6} P_\mathrm{M} \mathbb{E}\left[G_i\right] \frac{1}{2\pi} \int_{-\pi}^{\pi} \ell(d_{r,\Delta}) d\phi'. \qquad (3.7)$$

The result is obtained by plugging (3.3) into (3.6), exchanging sum and integral, and exploiting the linearity of the expectation. It is equivalent to $I_C(r)$ in (3.4) and, consequently, also yields (3.5). Thus, the result is *independent of the user's angle-position*. It can be interpreted as the *average expected* interference, i.e., the interference experienced by a *typical* user in a hexagonal scenario at distance r.

From (3.5) it is observed that the average expected interference increases by either increasing the transmit power P_M, decreasing the distance of the interferers R, or moving the user further away from the origin, which is reflected by the parameter r. The fading enters the equation only via the expectation, i.e, (3.4) and (3.7) hold for arbitrary fading distributions with finite mean. Finally, note that the circular interference model is not restricted to hexagons. By replacing '6' by 'N' in (3.3)–(3.7), it can generally be applied for substituting any convex regular N-polygonal model, as validated in Section 3.4.1.

3.3. Statistics of Aggregate Interference

In this section, aggregate interference in a hexagonal scenario with i.i.d. Gamma fading is investigated. Motivated by the findings in Section 2.2.3, it is proposed to approximate its statistics by a *single Gamma RV*. The corresponding shape- and scale parameters are dependent on the distance and can be determined by applying the previously presented circular model.

3.3.1. Interference Statistics at the Center

Assume i.i.d. Gamma fading with $G_i \sim \Gamma\left[k_0, \theta_0\right]$. Then, according to Section 3.2, interference can be considered as a sum of Gamma RVs, which are weighted by the factors $P_\mathrm{M} \ell(d_{r,\Delta_i}^{(6)})$, i.e., the received power without fading.

At the center of a hexagonal scenario (i.e., at $r = 0$), all weighting factors are equal, i.e., $P_\mathrm{M} \ell(d_{r,\Delta_i}^{(6)}) = P_\mathrm{M} \ell(R)$. By virtue of the scaling- and summation property of a Gamma RV

Chapter 3. Modeling Regular Aggregate Interference by Symmetric Structures

(conf. Section 2.2.3), the resulting interference is distributed as

$$I_6(0,\phi) \sim \Gamma[6\,k_0, \theta_0 P_M \ell(R)]. \tag{3.8}$$

3.3.2. Interference Statistics outside the Center

Outside the center (i.e., at $r > 0$), the distances $d_{r,\Delta_i}^{(6)}$ and, thus, also the weighting factors $P_M \ell(d_{r,\Delta_i}^{(6)})$ generally differ from each other. Hence, a non-uniform impact of the interferers is observed. Then, the interference statistics are only accessible via evaluating the distribution of a sum of Gamma RVs with varying scale parameter. This method is particularized in Chapter 4.

The current chapter resorts to the following first order estimate. It is proposed to approximate the typically experienced interference distribution at distance r, $0 < r \leq R/2$, by

$$\hat{I}(r) \sim \Gamma[\hat{k}(r), \hat{\theta}(r)]. \tag{3.9}$$

The rationale for this model are findings in prior work, where out-of-cell interference in stochastic networks is appropriately assessed by a Gamma distribution [17]. If it can be proven as accurate, it considerably facilitates further performance analysis by applying the methods in Section 2.2.3.

The distribution in (3.9) is fully determined by the distance-dependent shape- and scale parameters $\hat{k}(r)$ and $\hat{\theta}(r)$, respectively. In order to evaluate the two parameters, firstly the proposed circular interference model is employed to determine expectation and variance of $\hat{I}(r)$. Then, it is exploited that $\mathbb{E}[\hat{I}(r)] = \hat{k}(r)\,\hat{\theta}(r)$ and $\text{Var}[\hat{I}(r)] = \hat{k}(r)\,\hat{\theta}^2(r)$.

As discussed in Section 3.2.1, the distinct received powers from the interfering BSs can be *averaged* along a circle of radius r. Thus, the typical impact of one interferer is calculated as

$$P_M \frac{1}{2\pi} \int_{-\pi}^{\pi} \ell(d_{r,\Delta})\,d\Delta = \frac{P_M}{c_P} \frac{1}{R^2 - r^2}, \tag{3.10}$$

and yields the average expected interference at distance r as

$$\mathbb{E}[\hat{I}(r)] = 6\,k_0\theta_0 \frac{P_M}{c_P} \frac{1}{R^2 - r^2}. \tag{3.11}$$

The variance of the aggregate interference comprises two components:

1. The variance of the fading, which calculates as

$$\text{Var}_f\left[\hat{I}(r)\right] = 6k_0 \left(\theta_0 \frac{P_M}{c_P} \frac{1}{R^2 - r^2}\right)^2. \tag{3.12}$$

2. The variance of the received power without fading, which is caused by the unequal distances d_{r,Δ_i}. With

$$\frac{1}{2\pi} \int\limits_{-\pi}^{\pi} (P_M \ell(d_{r,\Delta}) - P_M \ell(R))^2 \, d\Delta = \left(\frac{P_M}{c_P R^2}\right)^2 \frac{2r^2 R^4 + r^4 R^2 - r^6}{(R^2 - r^2)^3}, \tag{3.13}$$

the second variance component is obtained as

$$\text{Var}_d\left[\hat{I}(r)\right] = 6k_0 \left(\theta_0 \frac{P_M}{c_P R^2}\right)^2 \frac{2r^2 R^4 + r^4 R^2 - r^6}{(R^2 - r^2)^3}. \tag{3.14}$$

Since the two components are statistically independent, the overall variance is calculated as

$$\begin{aligned}\text{Var}\left[\hat{I}(r)\right] &= \text{Var}_f\left[\hat{I}(r)\right] + \text{Var}_d\left[\hat{I}(r)\right] \\ &= 6k_0 \left(\theta_0 \frac{P_M}{c_P} \frac{1}{R^2 - r^2}\right)^2 \left(1 + \frac{2r^2 R^4 + r^4 R^2 - r^6}{R^6 - r^2 R^4}\right)\end{aligned} \tag{3.15}$$

where $\text{Var}_f[\hat{I}(r)]$ and $\text{Var}_d[\hat{I}(r)]$ refer to (3.12) and (3.14), respectively.

Finally, the distance-dependent shape- and scale parameter are derived from (3.11) and (3.15) as

$$\hat{k}(r) = 6k_0 \frac{R^4(R^2 - r^2)}{R^6 + r^2 R^4 + r^4 R^2 - r^6}, \tag{3.16}$$

$$\hat{\theta}(r) = \theta_0 \frac{P_M}{c_P} \frac{1}{R^2 - r^2} \left(1 + \frac{2r^2 R^4 + r^4 R^2 - r^6}{R^6 - r^2 R^4}\right). \tag{3.17}$$

3.4. Numerical Results and Discussion

In this section, the accuracy of the circular model and the proposed Gamma approximation are verified by numerical evaluation.

Chapter 3. Modeling Regular Aggregate Interference by Symmetric Structures

Table 3.1.: System parameters for validation. Transmit power and circle radius are referred from [167].

Parameter	Value
Transmit power	$P_M = 40\,\text{W}$
Circle radius	$R = 500\,\text{m}$
Path loss intercept	$b_P = 1$
Path loss constant	$c_P = 1$
Path loss exponent	$\alpha = 2$
Fading distribution	$G_i \sim \Gamma[1,1]$

3.4.1. Validation of Expected Aggregate Interference

First, the expected interference powers in the hexagonal reference scenario and the proposed circular interference setup are compared to each other. The transmit power and inter-site distance are specified as $P_M = 40\,\text{W}$ and $R = 500\,\text{m}$, based on the standard 3rd Generation Partnership Project (3GPP) macro cell scenario from [167]. Intercept and constant of the path loss $\ell(\cdot)$ are set $b_P = 1$ and $c_P = 1$ for simplicity. Fading is assumed to be distributed as $G_i \sim \Gamma[1,1]$. The parameters are summarized in Table 3.1.

Consider a user which is moved along a semi circle $\{(r,\phi)|\phi \in [0,\pi]\}$, as indicated in Figure 3.1. The expected interference in the hexagonal scenario is calculated as

$$\mathbb{E}\left[I_6(r,\phi)\right] = \sum_{i=1}^{6} P_M\, \mathbb{E}[G_i]\, \ell\left(d^{(6)}_{r,\Delta_i}\right), \qquad (3.18)$$

with $I_6(r,\phi)$ from (3.3) and $\mathbb{E}[G_i] = 1$. For the circular model, $\mathbb{E}\left[I_C(r)\right] = I_C(r)$, with $I_C(r)$ from (3.5). Figure 3.2 depicts the evaluated results of (3.5) and (3.18) for various distances r. It is observed that

- At cell-center, i.e., at $r = 0\,\text{m}$, the expected interference powers in the hexagonal- and circular scenario ($\mathbb{E}[I_6(0,\phi)]$ and $I_C(0)$) are equal.

- Outside the center, i.e., at $r > 0\,\text{m}$, $\mathbb{E}[I_6(r,\phi)]$ fluctuates around $I_C(r)$. The deviation is weak in the middle of the cell ($r = 125\,\text{m}$), and strong at cell-edge ($r = 250\,\text{m}$). Note that $\mathbb{E}[I_6(r,\phi)]$ is not symmetric about $I_C(r)$ due to the concavity of the path loss model.

The relative error of the circular interference model is calculated as

$$\epsilon(r,\phi) = \left|\frac{\mathbb{E}\left[I_6(r,\phi)\right] - I_C(r)}{\mathbb{E}\left[I_6(r,\phi)\right]}\right|, \qquad (3.19)$$

3.4. Numerical Results and Discussion

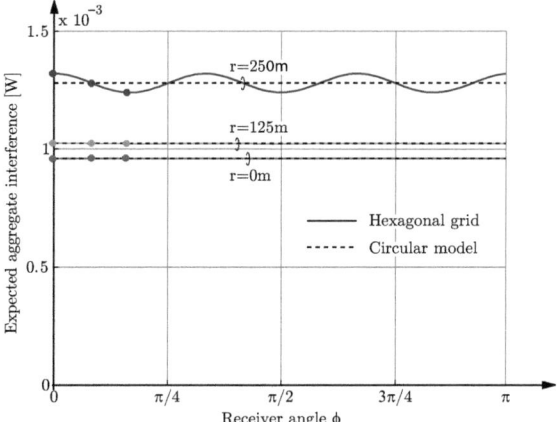

Figure 3.2.: Expected aggregate interference experienced at position (r,ϕ) in circular- ($I_C(r)$) and hexagonal model ($\mathbb{E}[I_6(r,\phi)]$), respectively. Receiver distances $r = \{0, 125, 250\}$ m refer to cell-center, middle of cell and cell-edge, respectively.

with $I_C(0)$ and $\mathbb{E}[I_6(0,\phi)]$ from (3.5) and (3.18), respectively. The largest error occurs at cell-edge, i.e.,

$$\max_{r,\phi} \epsilon(r,\phi) = \max_{\phi} \epsilon(250,\phi). \qquad (3.20)$$

In the specified scenario, $\max_\phi \epsilon(250,\phi) = 3.2\,\%$, as shown in Figure 3.3.

3.4.2. Validation of Gamma Approximation

In this section, the accuracy of the Gamma approximation in (3.9) and its parameterization by (3.16) and (3.17) are verified. The exact position-dependent distributions of $I_6(r,\phi)$ are obtained by evaluating Theorem 4.1.

In order to capture a *representative* profile of distributions, three user distances $r = \{0, 125, 250\}$ m and three angle-positions $\phi = \{0, \frac{\pi}{12}, \frac{\pi}{6}\}$ are considered, as illustrated by bold dots in Figure 3.4. The distances correspond to cell-center, middle of cell and cell-edge, respectively. The angle $\phi = 0$ represents a user, which is moved directly towards its strongest interferer, $\phi = \frac{\pi}{6}$ refers to the path with two equidistant dominant interferers, and $\phi = \frac{\pi}{12}$ is a variation thereof.

Chapter 3. Modeling Regular Aggregate Interference by Symmetric Structures

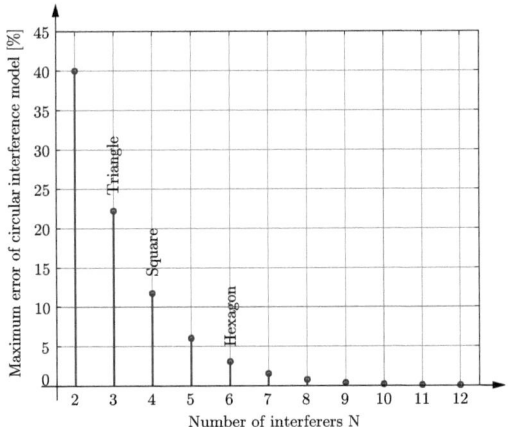

Figure 3.3.: Maximum deviation of circular interference model from expected interference in *convex regular N-polygonal* models. The labeled cell-shapes can be arranged in a grid without overlapping areas.

Fading is specified as $G_i \sim \Gamma[2,1]$. This corresponds to a 1×2 Single-Input-Multiple-Output (SIMO) system with Rayleigh-fading and MRC at the user, or, equivalently, a 2×1 Multiple-Input-Single-Output (MISO) system with MRT at the BS.

The CDF of the Gamma approximation, $F_{\hat{f}}(x; \hat{k}(r), \hat{\theta}(r))$ and the CDF $F_6(x; r, \phi)$ of $I_6(r, \theta)$ are evaluated at each distance r and angle ϕ, respectively. Referring to Section 2.2.2, the accuracy of the Gamma approximation is *quantified* by KS statistics, which formulate as

$$D_{\text{KS}}(r, \phi_m) = \sup_x \left| F_{\hat{f}}\left(x; \hat{k}(r), \hat{\theta}(r)\right) - F_6\left(x; r, \phi\right) \right|. \qquad (3.21)$$

Results are depicted in Figure 3.5. The Gamma approximation most closely resembles the experienced interference distributions at $\phi = \frac{\pi}{12}$. In this case, the difference between exact- and approximated CDFs is less than 1% for $r < 159\,\text{m}$ and 2.75% at cell-edge ($r = 250\,\text{m}$). The largest deviation occurs at $\phi = \frac{\pi}{6}$, where the user is moved centrally in between its two dominant interferers (upper curve). Then, the distributions differ by less than 1% for $r < 155\,\text{m}$ and by 3.7% at cell-edge.

For *qualitative* evaluation, Figure 3.6 depicts the exact CDFs and the corresponding Gamma approximations at the specified representative user positions, which are denoted by bold dots in

3.4. Numerical Results and Discussion

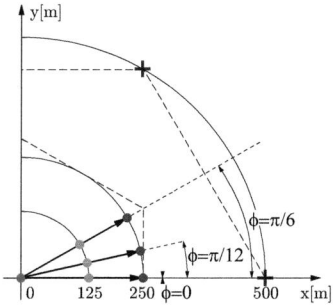

Figure 3.4.: Setup for evaluation. Cutout of Figure 3.1 (upper right quadrant).

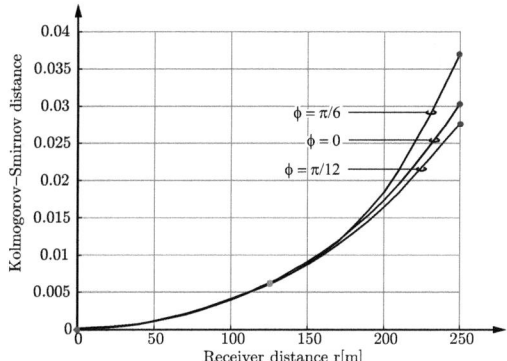

Figure 3.5.: KS statistics at position (r, ϕ_m), comparing Gamma approximation and exact distribution. Receiver distances $r = \{0, 125, 250\}$ m refer to cell-center, middle of cell and cell-edge, respectively.

27

Chapter 3. Modeling Regular Aggregate Interference by Symmetric Structures

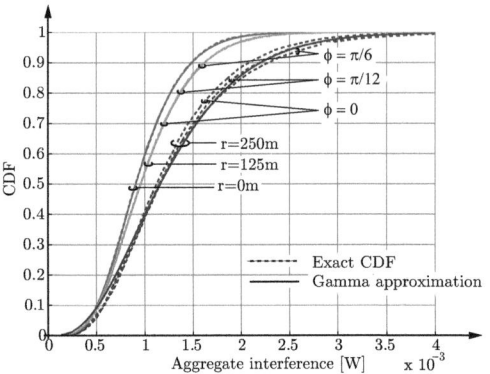

Figure 3.6.: Aggregate interference at particular user positions (see bold dots in Figure 3.4): Exact CDFs, as obtained by numerically evaluating [101] for a hexagon scenario (dashed lines) and corresponding Gamma approximations (solid lines).

Figures 3.4 and 3.5, respectively. The Gamma CDFs perfectly fit at cell center ($r = 0$ m) and in the middle of the cell ($r = 125$ m). At cell-edge ($r = 250$ m), the Gamma approximation closely resembles the experienced interference of a user at $\phi = \frac{\pi}{12}$. The probability of high interference values at $\phi = \frac{\pi}{6}$ is slightly underestimated by at most 3.7 % (conf. Figure 3.5).

3.5. Application in Heterogeneous Networks

In this section, aggregate interference statistics in a two-tier heterogeneous cellular network with regularly placed macro-BSs and randomly distributed small cell BSs are investigated. The interference contribution from each tier is approximated by a single Gamma RV and the total interference is calculated as the sum of the two. The accuracy of the approximations is verified by extensive Monte Carlo simulations.

The macro-tier comprises six hexagonally arranged BSs at distance $R = 500$ m, each transmitting with $P_M = 40$ W. Small cell BSs are distributed according to a PPP of density $\mu_S = 10^{-4}$ m^{-2} and transmit with a power of $P_S = 0.4$ W. As indicated in Figure 3.7, they are excluded from a ball[1] of radius $R_{Ex} = (P_M/P_S)^{-1/\alpha} R/2$ around the user so as to ensure user association to the central macro-BS at cell-edge. In both tiers, the path loss $\ell(\cdot)$ is modeled according to (3.1),

---
[1] In fact, it is a disc, but *ball* is the more common term in literature, e.g., in [17].

3.5. Application in Heterogeneous Networks

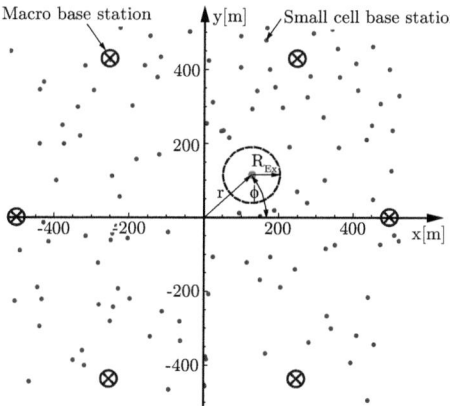

Figure 3.7.: Snapshot of a heterogeneous network deployment. Macro-BSs are arranged on a hexagon. Small cell BSs are randomly distributed around a user at (r, ϕ) and excluded from a ball of radius R_{Ex}.

with intercept $b_\text{P} = 1$, constant $c_\text{P} = 1$ and exponent $\alpha = 4$. Fading is assumed to be distributed as $G_i \sim \Gamma[1,1]$. The parameters are summarized in Table 3.2.

In the first step, the interference contribution from the macro-tier is approximated by a Gamma RV $\hat{I}_\text{M}(r) \sim \Gamma[\hat{k}_\text{M}(r), \hat{\theta}_\text{M}(r)]$. According to Section 3.3, it can be parameterized by the circular

Table 3.2.: Parameters for numerical evaluation of heterogeneous scenario.

Parameter	Value
Macro-BS transmit power	$P_\text{M} = 40\,\text{W}$
Inter macro-site distance	$R = 500\,\text{m}$
Small cell BS transmit power	$P_\text{S} = 0.4\,\text{W}$
Small cell density	$\mu_\text{S} = 10^{-4}\,\text{m}^{-2}$
Path loss intercept	$b_\text{P} = 1$
Path loss constant	$c_\text{P} = 1$
Path loss exponent	$\alpha = 4$
Fading distribution	$G_i \sim \Gamma[1,1]$

Chapter 3. Modeling Regular Aggregate Interference by Symmetric Structures

interference model. Recalculating (3.11), (3.12), (3.14), and (3.15) for $\alpha = 4$ yields

$$\hat{k}_{\mathrm{M}}(r) = \frac{6\,k_0 R^8 \left(r^2 - R^2\right)\left(r^2 + R^2\right)^2}{r^{14} - 7r^{12}R^2 + 23r^{10}R^4 - 41r^8 R^6 + 39r^6 R^8 - 25r^4 R^{10} - 9r^2 R^{12} - R^{14}}, \quad (3.22)$$

$$\hat{\theta}_{\mathrm{M}}(r) = \frac{\theta_0 P_{\mathrm{M}} \left(-r^{14} + 7r^{12}R^2 - 23r^{10}R^4 + 41r^8 R^6 - 39r^6 R^8 + 25r^4 R^{10} + 9r^2 R^{12} + R^{14}\right)}{c_{\mathrm{P}}\left(r^2 - R^2\right)^4 \left(r^2 + R^2\right) R^8}. \quad (3.23)$$

Secondly, the contribution of the small cell tier is also approximated by a Gamma RV $\hat{I}_{\mathrm{S}} \sim \Gamma[\hat{k}_{\mathrm{S}}, \hat{\theta}_{\mathrm{S}}]$. Along the lines of [26, Eqs. (2.19) and (2.21)], mean and variance of the actual interference $I_{\mathrm{A,S}}$ from the PPP model are determined as

$$\mathbb{E}[I_{\mathrm{A,S}}] = P_{\mathrm{S}}\,\mathbb{E}[G_i]\,2\pi\mu_{\mathrm{S}} \int_{R_{\mathrm{Ex}}}^{\infty} r\,\ell(r)\,dr$$

$$= P_{\mathrm{S}}\,k_0\,\theta_0 \frac{1}{c_{\mathrm{P}}} \pi\mu_{\mathrm{S}}\, R_{\mathrm{Ex}}^{-2}, \quad (3.24)$$

$$\mathrm{Var}[I_{\mathrm{A,S}}] = \mathbb{E}[G_i^2]\,2\pi\mu_{\mathrm{S}} \int_{R_{\mathrm{Ex}}}^{\infty} r\,\ell(r)^2\,dr$$

$$= (1 + k_0)\,k_0\,\theta_0^2\,P_{\mathrm{S}}^2\,\frac{1}{c_{\mathrm{P}}^2}\,\frac{\mu_{\mathrm{S}}\pi}{3}\,R_{\mathrm{Ex}}^{-6}. \quad (3.25)$$

Then, exploiting the identities $\mathbb{E}[\hat{I}_{\mathrm{S}}] = \hat{k}_{\mathrm{S}}\,\hat{\theta}_{\mathrm{S}}$ and $\mathrm{Var}[\hat{I}_{\mathrm{S}}] = \hat{k}_{\mathrm{S}}\,\hat{\theta}_{\mathrm{S}}^2$ yields

$$\hat{k}_{\mathrm{S}} = \frac{3 R_{\mathrm{Ex}}^2\,k_0\,\theta_0\,\mu_{\mathrm{S}}}{(1 + k_0)}, \quad (3.26)$$

$$\hat{\theta}_{\mathrm{S}} = \frac{P_{\mathrm{S}}(1 + k_0)}{3\,c_{\mathrm{P}}\,R_{\mathrm{Ex}}^{\alpha}}. \quad (3.27)$$

Finally, the PDF of the total aggregate interference, $\hat{I}_{\mathrm{A}}(r) = \hat{I}_{\mathrm{M}}(r) + \hat{I}_{\mathrm{S}}$, at user distance r is calculated as

$$f_{\hat{I}_{\mathrm{A}}}(x;r) = \hat{\theta}_{\mathrm{M}}(r)^{-\hat{k}_{\mathrm{M}}(r)}\,\hat{\theta}_{\mathrm{S}}^{-\hat{k}_{\mathrm{S}}}\,e^{-\frac{x}{\hat{\theta}_{\mathrm{S}}}}\,x^{\hat{k}_{\mathrm{M}}(r)+\hat{k}_{\mathrm{S}}-1}$$

$$\times\,{}_1\tilde{F}_1\!\left(\hat{k}_{\mathrm{M}}(r);\hat{k}_{\mathrm{M}}(r)+\hat{k}_{\mathrm{S}};\left(\frac{1}{\hat{\theta}_{\mathrm{S}}} - \frac{1}{\hat{\theta}_{\mathrm{M}}(r)}\right)x\right), \quad (3.28)$$

where ${}_1\tilde{F}_1(\cdot;\cdot;\cdot)$ denotes the regularized confluent hypergeometric function.

In order to verify the accuracy of this approximation, Monte Carlo simulations are carried out. The results for a *typical* user at distance r are obtained by averaging over 10^6 uniformly distributed angle-positions on $[0, 2\pi]$. For each position, 10^5 fading- and 10^4 spatial realizations of the small cell deployment are generated. The small cell BSs are distributed over a circular area of radius $10\,\text{R}$.

Figure 3.8(a) depicts the individual interference contributions from the macro- and the small cell tier at various user distances r. It is observed that the CDFs for the macro tier, which correspond to the approximation in (3.22) and (3.23), show an accurate fit with the Monte Carlo simulations. This corroborates the claim in Section 3.2.1 that the circular model is also applicable for path loss exponents other than $\alpha = 2$. The interference CDF of the small cell tier, which refers to the approximation in (3.26) and (3.27), is independent of the user distance r due to the fixed exclusion radius R_{Ex}. It is also in close agreement with the simulations. Figure 3.8(b) shows the CDFs of the aggregate interference from both macro- and small cell tier. It is found that the approximation by a sum of two independently parameterized Gamma RVs almost perfectly captures the actual interference characteristics at the cell center ($r = 0\,\text{m}$) and in the middle of the cell ($r = 125\,\text{m}$). It even provides an accurate fit at cell-edge ($r = 250\,\text{m}$).

3.6. Summary

In this chapter, a circular interference model for aggregate interference analysis in regular grid deployments is introduced. Particular focus is placed on characterizing a user at an eccentric location. The expected interference from the circular model is identified as the interference that is experienced by a typical user in a hexagonal grid at a certain distance from the origin. At cell-edge, it deviates by at most $3.2\,\%$ from the actual values.

In a second step, the corresponding interference statistics are approximated by a single Gamma RV. By means of the circular model, the distance-dependent shape- and scale parameters are determined in closed form and unveil the two key formative components of the distribution as the variance of the fading and the variance of the path loss due to the eccentric user location, respectively. Qualitative- and quantitative comparisons with the exact distributions confirm the accuracy of the Gamma approximation, yielding KS statistics no higher than $3.7\,\%$.

The chapter is completed by demonstrating the circular model's expedient adaption for representing the *well-planned part* of a two-tier heterogeneous cellular network. The example merges a fully regular macro-deployment with completely randomly distributed small cells and models the interference contribution from each tier by a single Gamma RV. The resulting aggregate interference distribution shows a remarkably good fit with Monte Carlo simulations. Hence, the

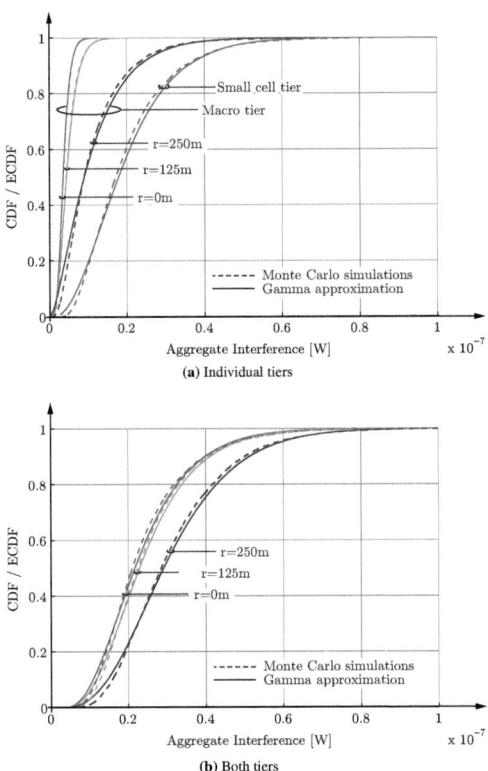

Figure 3.8.: CDFs of interference from macro- and small cell tier (a) and aggregate interference from both tiers (b). Solid lines indicate results as obtained by approximating the contribution of each tier by a Gamma RV. Dashed lines show results from Monte Carlo simulations. User distances $r = \{0, 125, 250\}$ m refer to cell-center, middle of cell and cell-edge, respectively.

3.6. Summary

model enables to accurately capture the impact of *both* user eccentricity *and* heterogeneity of the network with only few key parameters.

In a self-critical retrospection, the following aspects may be worth to revise. The approximation of the aggregate interference distribution by a single Gamma RV strongly relies on the assumption of Gamma fading, which, by itself, is often times a simplification of the actual fading distribution, as detailed in Chapter 6. Thus, it should be viewed as a simple yet accurate first order estimate. Furthermore, there is no known method to prove whether the Gamma distribution is the most suitable statistic for the aggregate interference in the presented scenario. Keeping an eye on analytical tractability, other two-parameter distributions might yield more accurate results.

It is disputable whether the hexagonal grid model is sustainable in future heterogeneous cellular networks. As of this writing, different spatial stochastic approaches such as the Ginibre- and the Poisson hardcore process may offer a more promising solution for modeling the planned part of the network deployment [46, 61, 69, 115]. Lastly, the presented circular model does not allow to account for power control and coordination among BSs. This is a major motivation for the next chapter, which extends the model by non-uniform power profiles.

Chapter 4.

Modeling Asymmetric Aggregate Interference by Symmetric Structures

In this chapter, the circular model from Chapter 3 is extended by non-uniform power profiles along the circles. The enhanced model enables to aggregate given interferer deployments such that the original interference statistics are accurately preserved while the amount of relevant interferers is reduced considerably.

Scaling up the number of base stations per unit area is one of the major trends in mobile cellular systems of the fourth (4G)- and fifth generation (5G) [169], making it increasingly difficult to characterize aggregate interference statistics with system models of low complexity. Tractable interference statistics have mainly been reported in the field of stochastic geometry. However, when closed-form expressions are desired, this mathematical framework imposes its own particular limitations, typically including spatial stationarity and isotropy of the scenario [14, 19, 47]. Hence, the potential to consider an asymmetric interference impact is very limited and notions such as *cell-center* and *cell-edge* are, in general, not accessible. Based on [77], the contributions of this chapter outline as follows:

- A new *circular interference model* is introduced. The key idea is to map arbitrary *out-of-cell* interferer deployments onto circles of uniformly spaced nodes such that the original aggregate interference statistics can accurately be reproduced. The model greatly reduces complexity as the number of participating interferers is significantly reduced.

- A *mapping scheme* that specifies a procedure for determining the power profiles of arbitrary interferer deployments is proposed. Its performance is evaluated by means of KS statistics. The test scenarios are modeled by PPPs so as to confront the regular circular structure with complete spatial randomness. It is shown that the individual spatial realizations exhibit largely diverging power profiles.

Chapter 4. Modeling Asymmetric Aggregate Interference by Symmetric Structures

- A new finite sum representation for the PDF of the *sum of Gamma RVs with integer-valued shape parameter* is introduced to further enhance and validate interference analysis with the circular model. Its restriction to integer-valued shape parameters is driven by relevant use cases for wireless communication engineering and the availability of *exact* solutions. The key strength of the proposed approach lies in the ability to decompose the interference distribution into the contributions of the individual interferers.

- Statistics of aggregate interference with *asymmetric interference impact* are investigated. The asymmetry is induced by eccentrically placing a user in a generic, isotropic scenario. This setup is achieved by applying the introduced circular model with uniform power profiles. The model enables to employ the proposed finite sum representation. It is shown that the partition of the interference distribution is particularly useful to identify candidate BSs for user-centric BS collaboration schemes. Moreover, the framework allows to predict the corresponding SIR- and rate statistics.

The main focus of this chapter is on downlink transmission in cellular networks. A comparable framework for the uplink is found in [170].

4.1. Circular Interference Model

Consider the serving BS to be located at the origin. The proposed circular interference model is composed of C concentric circles of interferers, as shown in Figure 4.1. On circle $c \in \{1, \ldots, C\}$ of radius R_c, N_c interfering nodes are spread out equidistantly. The interferer locations are expressed in terms of polar coordinates as $(R_c, \Psi_{c,n})$, where $\Psi_{c,n} = 2\pi n/N_c - \phi_c$, with $n \in \{1, \ldots, N_c\}$ and $\phi_c \in [0, 2\pi)$. Each node is unambiguously assigned to a tuple (c, n) and labeled as $\mathcal{T}_{c,n}$. The central BS is denoted as $\mathcal{T}_{0,0}$. Some of the interferers on the circles may also become serving nodes when BS collaboration schemes are applied, as will be shown later in Section 4.4.3.

The interferers on the circles do not necessarily represent real physical sources. As illustrated in Figure 4.2, they rather correspond to the N_c mapping points of an angle-dependent *power profile* $p_c[n]$, with $\sum_{n=1}^{N_c} p_c[n] = 1$. Exemplary profiles of a single circle are shown in Figure 4.3. Intuitively, $p_c[n]$ condenses the interferer characteristics of an annulus with inner radius R_c (R_{in} in case of $c = 1$) and possibly outer radius R_{c+1} (R_{out} in case of $c = C$) such that the circular model equivalently reproduces the original BS deployment in terms of interference statistics. This technique enables to represent substantially large networks by a *finite- and well-defined* constellation of nodes. By reducing the number of relevant interferers, it greatly

4.1. Circular Interference Model

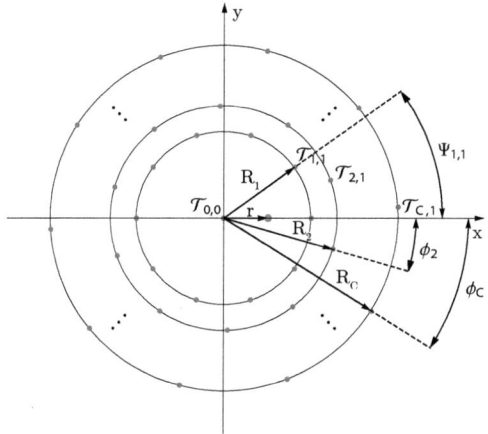

Figure 4.1.: Circular interference model with C circles of radius R_c and phase ϕ_c, $c \in \{1, \ldots, C\}$, and user at $(r, 0)$. $\mathcal{T}_{c,n}$ denotes the nodes of the model.

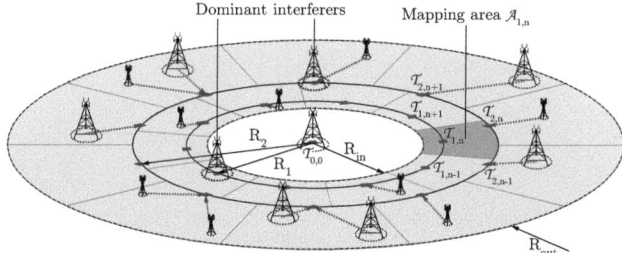

Figure 4.2.: Circular interference model with two circles, i.e., $C = 2$. Characteristics of an arbitrary heterogeneous interferer deployment are condensed to circles of equidistantly spaced nodes $\mathcal{T}_{c,n}$ such that the original interference statistics can accurately be reproduced. A mapping scheme is presented in Section 4.3. The original BSs are distributed within an annulus of inner radius R_in and outer radius R_out.

Chapter 4. Modeling Asymmetric Aggregate Interference by Symmetric Structures

Table 4.1.: Parameters of the circular interference model.

Symbol	Annotation
R_{in}	Inner radius of mapping region, $R_{\text{in}} \geq 0$
R_{out}	Outer radius of mapping region, $R_{\text{out}} > R_{\text{in}}$
C	Number of interferer circles, $C \in \mathbb{N}^+$
R_c	Radius of circle c, $c \in \{1,\ldots,C\}$, $R_c > 0$
ϕ_c	Phase of circle c, $c \in \{1,\ldots,C\}$ $\phi_c \in \left[-\frac{\pi}{N_c}, \frac{\pi}{N_c}\right]$
N_c	Number of mapping points, $c \in \{1,\ldots,C\}$, $N_c \in \mathbb{N}^+$
P_c	Total transmit power of circle c, $c \in \{1,\ldots,C\}$, $P_c > 0$
$p_c[n]$	Power profile of circle c, $c \in \{1,\ldots,C\}$, $n \in \{1,\ldots,N_c\}$, $p_c[n] \in [0,1]$

reduces complexity and thus allows to apply finite sum-representations as those introduced in Section 4.2.

Table 4.1 summarizes the parameters of the model. Typically, the size of the mapping region, as specified by R_{in} and R_{out}, is predetermined by the scenario. The freely selectable variables are the amount of circles C and, for each circle, the phase ϕ_c, the radius R_c and the number of mapping points N_c, respectively. Section 4.3 presents systematic experiments to provide a reference for the parameter setting and proposes a mapping scheme to determine power profiles $p_c[n]$ and transmit powers P_c, respectively.

A signal from node $\mathcal{T}_{c,n}$, located at $(R_c, \Psi_{c,n})$, to a user at $(r,0)$ experiences *path loss* $\ell(d_{c,n}(r))$, where $d_{c,n}(r) = \sqrt{R_c^2 + r^2 - 2R_c r \cos(\Psi_{c,n})}$ (conf. Figure 4.1) and $\ell(\cdot)$ is an arbitrary distance-dependent path loss law, as well as *fading*, which is modeled by statistically independent RVs $G_{c,n}$. The received power from node $\mathcal{T}_{c,n}$ at position $(r,0)$ is determined as

$$P_{\text{Rx},c,n}(r) = P_c p_c[n] \ell(d_{c,n}(r)) G_{c,n}, \tag{4.1}$$

where P_c denotes the total transmit power of circle c. It is important to note that the term $P_{\text{Rx},c,n}(r)$ can be interpreted as a RV $G_{c,n}$, which is scaled by a factor of $P_c p_c[n] \ell(d_{c,n}(r))$.

The nodes employ omnidirectional antennas with unit antenna gain. Characteristics of antenna directivity are incorporated into the power profile. In general, the central cell will have an irregular shape that can be determined by Voronoi tessellation [17]. For simplicity, the *small ball* approximation from [17] is applied. A user is considered as *cell-edge user*, if it is located at the edge of the central Voronoi cell's inscribing ball. This approximation misses some poorly covered areas at the actual cell-edge with marginal loss of accuracy [17].

Let $\mathcal{S}$ and $\mathcal{I}$ denote the sets of nodes $\mathcal{T}_{c,n}$ corresponding to desired signal and interference,

4.1. Circular Interference Model

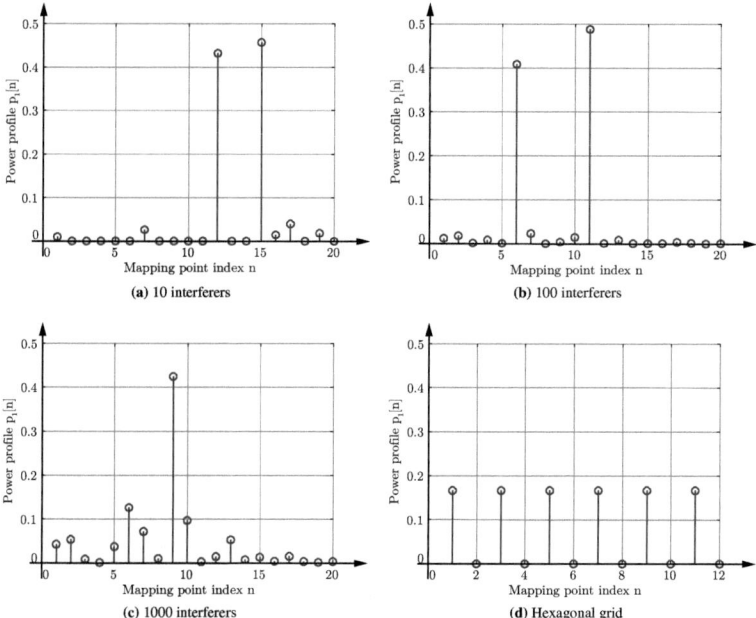

Figure 4.3.: Power profiles of circular models with one circle, i.e, $C = 1$, for three stochastic interference scenarios ((a)-(c)) with $N_1 = 20$ mapping points, and for a hexagonal grid with $N_1 = 12$ mapping points, respectively. The stochastic BS distributions are modeled by a PPP of intensity $\lambda = 10^{-6}\,\mathrm{m}^{-2}$. The expected number of interferers as denoted by the figure labels, is varied by altering the scenario size.

respectively. Then, the aggregate signal- and interference powers are calculated as

$$S(r) = \sum_{\{(c,n)|\mathcal{T}_{c,n}\in\mathcal{S}\}} P_{\text{Rx},c,n}(r), \qquad (4.2)$$

$$I(r) = \sum_{\{(c,n)|\mathcal{T}_{c,n}\in\mathcal{I}\}} P_{\text{Rx},c,n}(r), \qquad (4.3)$$

with $P_{\text{Rx},c,n}(r)$ from (4.1). The set $\mathcal{S}$ may include the central node $\mathcal{T}_{0,0}$ as well as nodes on the circles, if collaboration among the BSs is employed. The incoherence assumption is exploited for a more realistic assessment of the co-channel interference [171]. Following the interpretation of (4.1), (4.2) and (4.3) can be viewed as sums of scaled RVs, which are supported by a vast amount of literature for certain fading distributions such as Rayleigh, log-normal and Nakagami-m [89, 91–109].

This chapter places particular focus upon the Gamma distribution due to its wide range of useful features for wireless communication engineering. Preliminary information is provided in Section 2.2.3. The next section introduces a new theorem on the sum of Gamma RVs. The theorem is presented before validating the accuracy of the circular model as it is later exploited for this purpose.

4.2. Distribution of the Sum of Gamma Random Variables

As explained in Section 2.2.3, the Gamma distribution exhibits the *summation property*, i.e., if $G_i \sim \Gamma[k_i, \theta]$ with $i = 1, 2, \ldots, N$, then $\sum_{i=1}^{N} G_i \sim \Gamma[\sum_{i=1}^{N} k_i, \theta]$. While this feature is convenient to apply, it is the *sum of Gamma RVs* with *distinct* scale parameters that has attracted a lot of attention in describing wireless communications though. Most commonly, it emerged in the performance analysis of diversity combining receivers and the study of *aggregate co-channel interference under Rayleigh fading* [89, 91–100]. Therefore, communication engineers have considerably pushed the search for closed form statistics.

Representatively, Moschopoulos' much-cited series expansion in [101] was extended for correlated Gamma RVs in [89]. Other approaches based on the inverse Mellin transform (e.g., [172, 173]) paved the way for representations with a single integral as shown, e.g., in [92] or a Lauricella hypergeometric series as employed, e.g., in [91, 96].

All the aforementioned contributions focus on the sum of Gamma RVs with *real-valued* shape parameter. The resulting integrals and *infinite* series, despite being composed of elementary

4.2. Distribution of the Sum of Gamma Random Variables

functions, typically yield a slow rate of convergence. Therefore, an accurate approximation by a truncated series requires to keep a high amount of terms and complicates further analysis.

The sum of Gamma RVs with *integer* shape parameter has mainly been reported in statistical literature. Initial approaches focused on the moment generating function and results were obtained in the form of series expansions [102]. Based on the work of [103], [104] was among the first to formulate a convenient closed form solution. Soon after, the Generalized Integer Gamma (GIG) distribution was published in [105]. This approach was also adopted in wireless communication engineering [93, 95]. In comparison to RVs with *real-valued* shape parameter, the PDF of the sum of RVs with *integer* shape parameter allows an *exact* representation by a *finite* series.

4.2.1. Proposed Finite Sum Representation

In the analysis of aggregate interference statistics, it is particularly desirable to identify the main distribution-shaping factors, i.e., the interfering sources with the highest impact. However, the expressions in [93] and [95] are not suitable for this task due to multiple nested sums and recursions. The proposed finite-sum representation in this chapter avoids recursive functions and enables to *straightforwardly trace the main determinants* of the distribution characteristics.

Theorem 4.1. *Let $G_l \sim \Gamma[k_l, \theta_l]$ be L independent Gamma RVs with $k_l \in \mathbb{N}^+$ and all θ_l different[1]. Then, the PDF of $Y = G_1 + \cdots + G_L$ can be expressed as*

$$f_Y(y) = \sum_{l=1}^{L} \frac{\Lambda_l}{\theta_l^{k_l}} h_{k_l-1,l}(0) e^{-y/\theta_l} \tag{4.4}$$

with

$$\Lambda_l = \frac{(-1)^{k_l+1}}{(k_l-1)!} \prod_{i=1, i \neq l}^{L} \left(1 - \frac{\theta_i}{\theta_l}\right)^{-k_i}, \qquad l = 1, \ldots, L \tag{4.5}$$

$$h_{\delta+1,l}(\zeta) = h_{1,l}(\zeta) h_{\delta,l}(\zeta) + \frac{d}{d\zeta} h_{\delta,l}(\zeta), \qquad \delta = 0, \ldots, k_l - 1 \tag{4.6}$$

Chapter 4. Modeling Asymmetric Aggregate Interference by Symmetric Structures

and

$$h_{1,l}(0) = -y + \sum_{i=1, i \neq l}^{L} k_i \left(\frac{1}{\theta_i} - \frac{1}{\theta_l} \right)^{-1}, \quad l = 1, \ldots, L \quad (4.7)$$

$$h_{1,l}^{(m)}(0) = m! \sum_{i=1, i \neq l}^{L} k_i \left(\frac{1}{\theta_i} - \frac{1}{\theta_l} \right)^{-m-1}, \quad m = 1, \ldots, k_l - 1 \quad (4.8)$$

Proof. The proof is provided in Appendix C. □

Superscript (m) of $h_{1,l}^{(m)}(\zeta)$ denotes the m-th derivative of $h_{1,l}(\zeta)$. The recursive determination of $h_{\delta,l}(\zeta)$ in (4.6) seemingly interrupts the straightforward calculation of $f_Y(y)$. However, $h_{\delta,l}(\zeta)$ is a function of only $h_{1,l}(\zeta)$ and its higher order derivatives. Therefore, the function series in (4.6) can be evaluated *in advance* up to the highest required degree $\delta_{\max} = \max_l k_l - 1$.

Thus, the proposed approach enables the *exact* calculation of $f_Y(y)$ in a *component-wise* manner[2]. In the next step, it is shown how to apply Theorem 4.1 in the proposed circular model.

4.2.2. Application in Circular Interference Model

Assume that $G_{c,n} \sim \Gamma[k_{c,n}, \theta_{c,n}]$ in (4.1), with $k_{c,n} \in \mathbb{N}^+$ and $\theta_{c,n} > 0$. Then, (4.2) and (4.3) represent sums of scaled Gamma RVs $P_{\text{Rx},c,n}(r) \sim \Gamma[k_{c,n}, \theta'_{c,n}(r)]$, where $\theta'_{c,n}(r) = P_c p_c[n] \cdot \ell(d_{c,n}(r)) \theta_{c,n}$. Therefore, their PDFs can be determined by applying Theorem 4.1.

The theorem requires all scale parameters to be different. Thus, let $\boldsymbol{\theta_\mathcal{I}}(r)$ denote the vector of unique scale parameters $\theta'_{c,n}(r)$ with (c,n) from the set $\{(c,n)|\mathcal{T}_{c,n} \in \mathcal{I}\}$. A second vector $\mathbf{k_\mathcal{I}}$ contains the corresponding shape parameters. By virtue of the *summation property*, if $\theta'_{c,n}(r)$ occurs multiple times in the set, the respective shape parameter in $\mathbf{k_\mathcal{I}}$ is calculated as the sum

---
[1] The uniqueness of θ_l can be assumed without loss of generality. In case of some θ_l being equal, the corresponding RVs are added up by virtue of the *summation property of Gamma RVs* (conf. Section 2.2.3).

[2] A *Mathematica®* implementation is provided in Appendix D. The code is conveniently separated into the pre-calculation, storing and reloading of the auxiliary functions in (4.6), and the computation of the actual distribution function.

of shape parameters $k_{c,n}$ of the according entries. The vectors $\boldsymbol{\theta}_\mathcal{S}(r)$ and $\mathbf{k}_\mathcal{S}$ are obtained equivalently. Then, the PDFs of $S(r)$ and $I(r)$ are expressed as

$$f_S(\gamma;r) = \sum_{l=1}^{L_\mathcal{S}} \frac{\Lambda_l}{\theta_l(r)^{k_l}} h_{k_l-1,l}(0) e^{-\gamma/\theta_l(r)}, \qquad (4.9)$$

$$f_I(\gamma;r) = \sum_{l=1}^{L_\mathcal{I}} \frac{\Lambda_l}{\theta_l(r)^{k_l}} h_{k_l-1,l}(0) e^{-\gamma/\theta_l(r)}, \qquad (4.10)$$

with Λ_l and $h_{\delta,l}(\cdot)$ as defined in (4.5) and (4.6). Subscript l indicates the l-th components of the vectors $\mathbf{k}_\mathcal{S}$ ($\boldsymbol{\theta}_\mathcal{S}(r)$) and $\mathbf{k}_\mathcal{I}$ ($\boldsymbol{\theta}_\mathcal{I}(r)$) and $L_\mathcal{S}$ and $L_\mathcal{I}$ are their corresponding lengths, respectively.

Hence, employing Theorem 4.1 allows to evaluate the *exact* distributions of the aggregate signal- and interference from the circular model by *finite* sums. In the following section, this fact is exploited to verify the accuracy of the model.

4.3. Mapping Scheme for Stochastic Network Deployments

This section presents a procedure to determine the power profiles $p_c[n]$ and the corresponding powers P_c of the circular model for completely random interferer distributions. Then, systematic experiments are carried out to provide a reference for selecting the free variables C and N_c, respectively. The parameters R_c and ϕ_c are also specified by the procedure. The accuracy of the approximation is measured by means of the KS distance. According to Section 2.2.2, it is defined as

$$D_{\mathrm{KS}}(r) = \sup_x |F_{I,\mathrm{original}}(x;r) - F_{I,\mathrm{circular}}(x;r)|, \qquad (4.11)$$

where r refers to the user's eccentricity and $F_{I,\mathrm{original}}(x;r)$ and $F_{I,\mathrm{circular}}(x;r)$ denote the aggregate-interference CDFs[3] of the original deployment and the circular model, respectively. The corresponding PDFs are obtained by applying Theorem 4.1.

4.3.1. Mapping Procedure

Let $\mathcal{N}$ denote a (possibly heterogeneous) set of BSs that are arbitrarily distributed within an annulus $\mathcal{A}$ of inner radius R_{in} and outer radius R_{out}, as shown in Figure 4.2. Radius R_{out} as

---
[3]The CDF of a RV X with PDF X is determined as $F_X(x) = \int_{-\infty}^{x} f_X(x')dx'$.

Algorithm 1: Mapping procedure for circular model.

Data: number of circles C; nodes per circle N_c;
    original base station deployment $\mathcal{N}$;
    inner- and outer radius of mapping region $\mathcal{A}$: R_in and R_out;
Result: P_c, $p_c[n]$, R_c and ϕ_c for all $c \leq C$;
for $c = 1$ to C **do**
| determine R_c and ϕ_c based on the strongest interferer that has not yet been mapped;
end
for $c = 1$ to C **do**
| specify mapping region $\mathcal{A}_c$ with inner radius R_c and outer radius R_{c+1};
| **if** $c = 1$ **then** set inner radius of $\mathcal{A}_c$ to R_in; **end**
| **if** $c = C$ **then** set outer radius of $\mathcal{A}_c$ to R_out; **end**
| compute P_c and $p_c[n]$ for $\mathcal{A}_c$;
end
---

well as the number of nodes in $\mathcal{N}$ could be substantially large. Given a circular model with C circles and N_c nodes per circle, the parameters P_c, R_c and ϕ_c as well as the power profile $p_c[n]$ can be determined by Algorithm 1.

The presented procedure employs the origin as a reference point and therefore does *not* depend on the user location. The computation of P_c and $p_c[n]$ outlines as follows. Let $\mathcal{T}_{c,n}$ denote node n on circle c. Assume that its associated mapping area $\mathcal{A}_{c,n}$ is bounded by the circles of radius R_c and R_{c+1} (in the case of $c = 1$, the inner radius is set to R_in; for $c = C$ the outer radius is set to R_out) as well as the perpendicular bisectors of the two line segments $\overline{\mathcal{T}_{c,n}\mathcal{T}_{c,n-1}}$ and $\overline{\mathcal{T}_{c,n}\mathcal{T}_{c,n+1}}$, as illustrated in Figure 4.2. This yields an even division of circle c's mapping area $\mathcal{A}_c$, which can be formulated as $\mathcal{A}_c = \bigcup_{n\in\{1,...,N_c\}} \mathcal{A}_{c,n}$. The average received power at the origin from all considered BSs in $\mathcal{A}_c$ is calculated as

$$P_{\text{Rx},\mathcal{A}_c} = \sum_{i \in \mathcal{N} \cap \mathcal{A}_c} P_{\text{Tx},i}\,\ell(d_i)\mathbb{E}[G_i], \tag{4.12}$$

where $P_{\text{Tx},i}$, d_i and G_i correspond to transmit power, distance and experienced fading of interferer i, respectively. Then, the total transmit power P_c is obtained by mapping $P_{\text{Rx},\mathcal{A}_c}$ back on the circle, which formulates as $P_c = P_{\text{Rx},\mathcal{A}_c}\,\ell(R_c)^{-1}$. Hence, in this scheme the average received powers from the original deployment and the circular model are equivalent at the origin. The segmentation of $\mathcal{A}_c$ into areas $\mathcal{A}_{c,n}$ yields the corresponding power profile

$$p_c[n] = \frac{1}{P_{\text{Rx},\mathcal{A}_c}} \left(\sum_{i \in \mathcal{N} \cap \mathcal{A}_{c,n}} P_{\text{Tx},i}\,\ell(d_i) \right), \tag{4.13}$$

4.3. Mapping Scheme for Stochastic Network Deployments

Table 4.2.: System setup for evaluation.

Parameter	Value
Transmit power	$P_{T1} = 40\,\text{W}$ ($P_{T2} = 4\,\text{W}$)
Node density	$\lambda = \{0.5 \cdot 10^{-6}, 10^{-6}\}\,\text{m}^{-2}$ ($\lambda_2 = 0.5 \cdot 10^{-5}\,\text{m}^{-2}$)
Expected number of interferers	$N_\text{I} = \{100, 1000\}$
Path loss	$\ell(x) = \min(b_\text{P}, 1/c_\text{P}\, x^{-4})$, $b_\text{P} = 1$, $c_\text{P} = 1$, $x > 0$
Fading	$G_{c,n} \sim \Gamma[2, 1]$

with $P_{\text{Rx},\mathcal{A}_c}$ from (4.12).

In the presented procedure, the parameters R_c and ϕ_c are set such that the c-th dominant interferer coincides with a node on circle c, as illustrated in Figure 4.2. This ensures that $R_1 \geq R_\text{in}$ (in a heterogeneous network, as investigated in Section 4.3.3, non-dominant interferers between R_in and R_1 are mapped "back" on circle 1 by the receive-power dependent weighting in (4.13)) and $R_C \leq R_\text{out}$, and is especially suitable for completely random interferer distributions, as demonstrated in the next section. In fully regular scenarios, on the other hand, a circle comprises multiple, equally dominant nodes, making it expedient to specify R_c and ϕ_c according to the structure of the grid. For example, the circular model allows to *perfectly* represent a hexagonal grid setup, when the number of mapping points is set as a multiple of six. Then, the nodes on the circle coincide with the actual interferer locations. An exemplary power profile for $N_1 = 12$ is shown in Figure 4.3(d).

Algorithm 1 is one of many possible mapping approaches. It is a heuristic, based on my experience and observations and is thus *not claimed* to be optimal. Its refinement yields an interesting topic for further work. In the next two sections, systematic experiments in *completely random* scenarios are performed to provide a reference for setting C and N_c. For reasons of clarity, Section 4.3.2 is limited to *homogeneous* BS deployments. Heterogeneous setups are then evaluated in Section 4.3.3. It is refrained from stochastic scenarios with a certain degree of regularity, since measuring spatial inhomogeneity is itself an ongoing topic of research [43]. Completely random- and fully regular scenarios are considered as limiting cases, encompassing every conceivable practical deployment in between.

4.3.2. Performance Evaluation for Homogeneous Base Station Deployments

The *original* interferer deployment $\mathcal{N}$ is modeled by a PPP of intensity λ. Such process is considered *most challenging* for the regularly structured circular model, as it represents complete spatial randomness. Signal attenuation is modeled by a log-distance dependent path loss law

Chapter 4. Modeling Asymmetric Aggregate Interference by Symmetric Structures

$\ell(x) = \min(b_P, 1/c_P\, x^{-4})$, and Gamma fading with $k = 2$ and $\theta = 1$, referring to a 2×1 MISO setup and MRT. In this chapter, $b_P = 1$ and $c_P = 1$ for simplicity. The BSs transmit with power $P_{T1} = 40\,\text{W}$ and are distributed within an annular regions of inner radius $R_{\text{in}} = 500\,\text{m}$ and $R_{\text{out}} = \sqrt{N_I/(\pi\lambda) + R_{\text{in}}^2}$. Radius R_{in} ensures that the inscribing ball of the central cell has a minimum radius of $250\,\text{m}$, assuming that the central BS also transmits with P_{T1}. The outer radii R_{out} are chosen such that, on average, N_I BSs locations are generated within the corresponding annulus[4]. In order to cover a wide range of scenarios, $N_I = \{100, 1000\}$ and $\lambda = \{0.5 \cdot 10^{-6}, 10^{-6}\}\,\text{m}^{-2}$ are studied. The parameter settings are summarized in Table 4.2.

For each scenario snapshot, eight circular models with $C = \{1, 2, 3, 4\}$ and two distinct values of N_c are set up according to Section 4.3.1. In the case of $\lambda = 0.5 \cdot 10^{-6}\,\text{m}^{-2}$, $N_c = \{10, 20\}$ and, for $\lambda = 10^{-6}\,\text{m}^{-2}$, $N_c = \{20, 40\}$, respectively. Then, the *aggregate interference distributions* are determined. The distributions for the original interferer deployment are only obtained via simulations (by averaging over 1000 spatial realizations and 10 000 fading realizations), since the vast amount of nodes hampers the application of Theorem 4.1 due to complexity issues. On the other hand, the circular models comprise at most 43 *active* nodes and therefore enable to utilize the theorem. This number is obtained for $C = 4$ and $N_c = 40$, and stems from the fact that in a homogeneous BS deployment, the dominant interferers are also the closest ones. Therefore, the presented scheme only maps a single BS on each circle $c < C$, i.e., except for $c = C$ there is only one active node per circle.

Figure 4.4 depicts KS distances over the user eccentricity r. The first important observation is that the accuracy considerably improves with an increasing number of circles C. This mainly results from accurately capturing the first few dominant BSs that have the largest impact on the aggregate interference distribution, as later shown in Section 4.4. A second remarkable observation is that doubling the amount of nodes per circle from $N_c = 10$ to $N_c = 20$ for $\lambda = 0.5 \cdot 10^{-6}\,\text{m}^2$ (conf. Figures 4.4(a) and 4.4(b)), and from $N_c = 20$ to $N_c = 40$ for $\lambda = 10^{-6}\,\text{m}^2$ (conf. Figures 4.4(c) and 4.4(d)) does not achieve smaller KS distances, respectively. This result indicates that it is rather the number of circles C than the number of nodes per circle N_c that impacts the accuracy. As shown in the examples, good operating points for homogeneous macro-BS deployments are $N_c = 20$ and $C = 4$, independent of the deployment parameters. Lastly, it should be noted that the circular model allows to represent 1000 and more interferers by some 10 nodes with KS distances at the cell-edge not exceeding 0.05.

4.3. Mapping Scheme for Stochastic Network Deployments

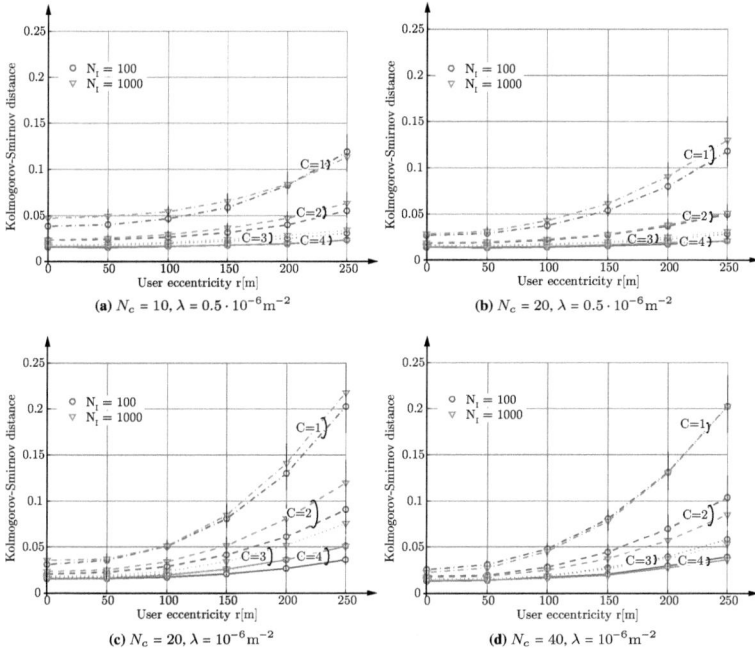

Figure 4.4.: KS distance over user eccentricity r. Plot markers $\{"\circ", "\triangledown"\}$ refer to various scenario sizes with $N_I = \{100, 1000\}$ expected interferers, respectively. Different line styles denote circular models with $C = \{1, 2, 3, 4\}$. Figure labels refer to the corresponding number of nodes per circle, N_c, and the spatial density λ of the original interferer deployment. Black bars depict 95% confidence intervals.

Chapter 4. Modeling Asymmetric Aggregate Interference by Symmetric Structures

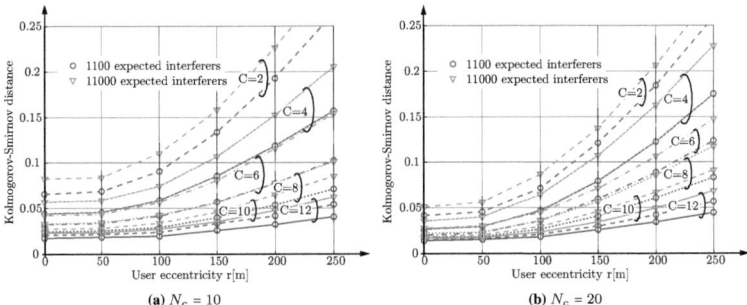

Figure 4.5.: KS distance over user eccentricity r for *heterogeneous* PPP scenarios with $\lambda = 0.5 \cdot 10^{-6}\,\mathrm{m}^{-2}$ ($P_{T1} = 40\,\mathrm{W}$) and $\lambda_2 = 0.5 \cdot 10^{-5}\,\mathrm{m}^{-2}$ ($P_{T2} = 4\,\mathrm{W}$). Plot markers $\{$" $\circ$ "," $\triangledown$ "$\}$ refer to various scenario sizes with $\{1100, 11\,000\}$ expected interferers, respectively. Different line styles denote circular models with $C = \{2, 4, 6, 8, 10, 12\}$. Figure labels refer to the corresponding number of nodes per circle N_c. Black bars depict 95% confidence intervals.

4.3.3. Performance Evaluation of Heterogeneous Base Station Deployments

In this section, a second independent PPP of intensity $\lambda_2 = 0.5 \cdot 10^{-5}\mathrm{m}^{-2}$ is added on top of the PPP scenarios with $\lambda = 0.5 \cdot 10^{-6}\mathrm{m}^{-2}$ in Section 4.3.2. The corresponding nodes transmit with normalized power $P_{T2} = 4\,\mathrm{W}$, thus representing a dense overlay of low power BSs. For simplicity, they are distributed within annuli of inner radius R_{in} and outer radii R_{out} as specified above[5]. This yields a total number of $\{1100, 11\,000\}$ expected interferers, respectively. For each snapshot, Algorithm 1 is applied with $C = \{2, 4, 6, 8, 10, 12\}$ and $N_c = \{10, 20\}$. The performance evaluation is carried out along the lines of Section 4.3.2 and the parameters are summarized in Table 4.2.

Figure 4.5 depicts the results in terms of KS distances. It is observed that, in accordance with Section 4.3.2, accuracy is rather improved by increasing the number of circles C than by employing more nodes per circle (i.e., increasing N_c). In the heterogeneous scenarios the number of circles has to be roughly tripled in order to achieve a performance similar to the

---

[4]Consider a PPP of intensity λ within an annulus of inner radius R_{in} and outer radius R_{out}. The expected number of generated nodes is calculated as $N_{\mathrm{I}} = \lambda(R_{\mathrm{out}}^2 - R_{\mathrm{in}}^2)\pi$.
[5]To ensure that the inscribing ball of the central cell has a radius of 250 m, an inner radius of $1 + (P_{T1}/P_{T2})^{-1/\alpha}$ would be sufficient.

4.3. Mapping Scheme for Stochastic Network Deployments

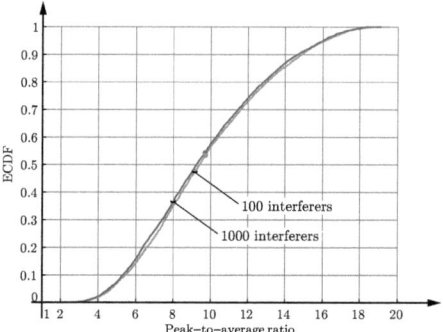

Figure 4.6.: Peak-to-average ratio of power profiles of PPP scenarios with intensity $\lambda = 10^{-6}\,\mathrm{m}^{-2}$ and $N_\mathrm{I} = \{100, 1000\}$ expected interferers. The corresponding circular models are obtained by Algorithm 1 with $C = 1$ and $N_1 = 20$. Bold dots denote the mean ratios.

homogeneous cases (conf. Figures 4.4(a) and 4.4(b)), although mapping 11 times as many interferers.

4.3.4. Power Profiles of PPP Snapshots

As indicated in Figure 4.3, power profiles of homogeneous PPP scenarios are characterized by one or a few large amplitudes. To quantify this claim, Figure 4.6 shows the empirical distributions of the power-profile peak-to-average ratios as obtained from the PPPs in Section 4.3.2 with $\lambda = 10^{-6}\,\mathrm{m}^{-2}$. The corresponding circular models encompass a single circle (i.e., $C = 1$) with $N_1 = 20$ mapping points. It is observed that the peak-to-average ratios range from 3 to 19 with the medians being located around 9.5. The presence of dominant interferers results in a large asymmetry of the interference impact. However, in modeling approaches that are based on stochastic geometry, the differences between scenarios at both ends of the scale are concealed by spatial averaging. What is more, such approaches commonly require user-centric isotropy of the setup in order to obtain exact solutions (e.g., circularly symmetric exclusion regions [17]). Hence, the differences between interference characteristics in the *center of the cell* and at *cell-edge* are generally not accessible. In the next section, the circular model is applied to generate a generic, circularly symmetric scenario and, by employing Theorem 4.1, analyzes the impact of user eccentricity.

Chapter 4. Modeling Asymmetric Aggregate Interference by Symmetric Structures

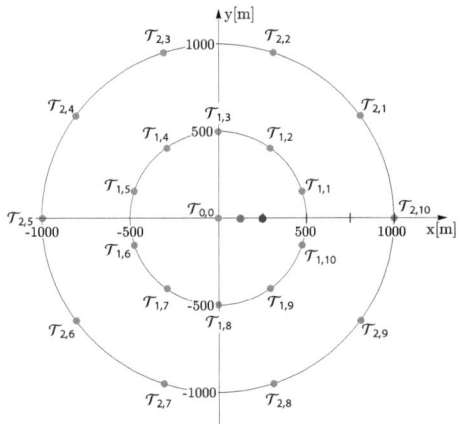

Figure 4.7.: Circular model with two two circles of radius $R_1 = 500\,\text{m}$ and $R_2 = 1000\,\text{m}$, respectively. Each circle employs 10 BSs. The BS positions are rotated by $\phi_1 = -\pi/10$ and $\phi_2 = 0$. Users at $(125\,\text{m}, 0)$ and $(250\,\text{m}, 0)$ are denoted as bold dots and refer to *middle of cell* and *cell-edge*, respectively.

4.4. Interference and Rate at Eccentric User Locations

In this section, user-centric BS collaboration schemes in scenarios with *asymmetric interferer impact* are investigated. The asymmetry can either arise from *non-uniform* power profiles or user locations outside the center of an otherwise isotropic scenario. The particular emphasis of this section is on the latter, since it is found less frequently in literature. In order to generate a generic, circularly symmetric scenario[6], the introduced circular model is applied, which enables to employ Theorem 4.1 for the analysis of the interference statistics.

4.4.1. Generic Circularly Symmetric Scenario

The network is composed of a central BS and two circles of interferers with $R_1 = 500\,\text{m}$ and $R_2 = 1000\,\text{m}$, as depicted in Figure 4.7. Each circle employs 10 interferers and a uniform power profile, i.e., $p_c[n] = 1/10$. The respective total transmit powers are specified as $P_1 = 400\,\text{W}$ and

---

[6]In fact, the circular model generates a *rotationally symmetric* scenario due to the finite number of nodes. However, by setting N_c sufficiently large, the scenario can be considered as *quasi-circularly symmetric*.

Table 4.3.: Parameters of circular model for numerical evaluation.

Circle	Parameters					
1	$R_1 = 500\,\text{m}$	$N_1 = 10$	$P_1 = 400\,\text{W}$	$\phi_1 = -\frac{\pi}{10}$	$p_1[n] = \frac{1}{10}$	$n \in \{1, \ldots, 10\}$
2	$R_2 = 1000\,\text{m}$	$N_2 = 10$	$P_2 = 800\,\text{W}$	$\phi_2 = 0$	$p_2[n] = \frac{1}{10}$	$n \in \{1, \ldots, 10\}$

$P_2 = 800\,\text{W}$, respectively. The interferer locations are assumed to be rotated by $\phi_1 = -\pi/10$ and $\phi_2 = 0$, respectively. BS $\mathcal{T}_{0,0}$ is located at the origin and $P_0 = 40\,\text{W}$.

The parameters of the circular model are summarized in Table 4.3 and the modeling of the signal propagation is referred from Table 4.2, respectively. The first goal is to identify the nodes, which dominate the interference statistics at eccentric user locations. Then, these insights are applied for user-centric BS coordination and -cooperation.

4.4.2. Components of Asymmetric Interference

In the first step, only the inner circle of interferers is assumed to be present, i.e., the set $\mathcal{I}$ comprises the 10 nodes $\mathcal{T}_{1,n}$, $n = 1, \ldots, 10$, of circle 1. The target is to determine the impact of the closest nodes on the aggregate interference statistics. For this purpose, two representative user locations at $r = R_1/4$ and $r = R_1/2$ are investigated, referring to *middle of cell* and *cell-edge*, respectively.

The PDF of the aggregate interference is obtained by Theorem 4.1. Its evaluation is simplified by the scenario's symmetry about the x-axis: (i) equal node-to-user distances from upper- and lower semicircle, i.e., $d_{1,n} = d_{1,10-n+1}$, (ii) uniform power profile $p_1(n) = 1/10$, and (iii) equal scale parameters $\theta_{1,n} = 1$. Thus, $\theta'_{1,n}(r) = \theta'_{1,10-n+1}(r)$, with $\theta'_{1n}(r) = P_1/10\,\ell(d_{1,n}(r))$. The vectors $\boldsymbol{\theta}_\mathcal{I}(r)$ and $\mathbf{k}_\mathcal{I}$ are of length $L^\mathcal{I} = 5$, with $[\boldsymbol{\theta}_\mathcal{I}(r)]_l = \theta'_{1,l}(r)$ and $[\mathbf{k}_\mathcal{I}]_l = 4$, respectively. Hence, the distribution of aggregate interference at distance r from the center formulates as

$$f_I(x;r) = \sum_{l=1}^{5} \frac{\Lambda_l}{\theta'_{1,l}(r)^4} h_{3,l}(0) e^{-x/\theta'_{1,l}(r)}, \tag{4.14}$$

where

$$\Lambda_l = -\frac{1}{6} \prod_{i=1, i \neq l}^{5} \left(1 - \frac{\theta_i}{\theta_l}\right)^{-4}, \quad l = 1, \ldots, 5, \tag{4.15}$$

$$h_{3,l} = (h_{1,l}(0))^3 + 3h_{1,l}(0)h_{1,l}^{(1)}(0) + h_{1,l}^{(2)}(0), \tag{4.16}$$

Chapter 4. Modeling Asymmetric Aggregate Interference by Symmetric Structures

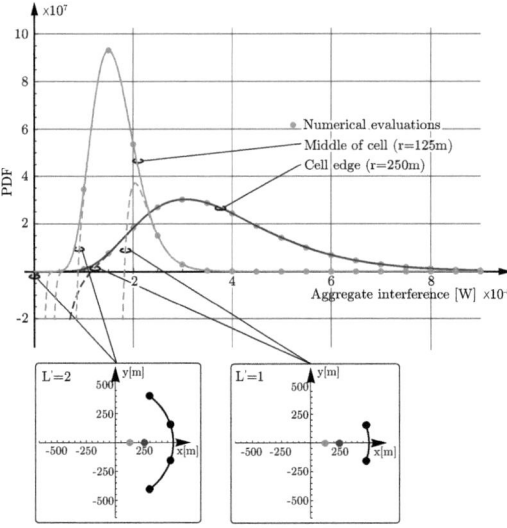

Figure 4.8.: Distribution of aggregate interference at user distances $r = 125\,\text{m}$ and $r = 250\,\text{m}$, respectively. Dots refer to results as obtained with the approach in [92]. Dashed curves show contribution from dominant interferers.

with

$$h_{1,l}(0) = -y + 4 \sum_{i=1,i\neq l}^{5} \left(\frac{1}{\theta_i} - \frac{1}{\theta_l}\right)^{-1}, \qquad (4.17)$$

$$h_{1,l}^{(1)}(0) = 4 \sum_{i=1,i\neq l}^{5} \left(\frac{1}{\theta_i} - \frac{1}{\theta_l}\right)^{-2}, \qquad (4.18)$$

$$h_{1,l}^{(2)}(0) = 8 \sum_{i=1,i\neq l}^{5} \left(\frac{1}{\theta_i} - \frac{1}{\theta_l}\right)^{-3}. \qquad (4.19)$$

Figure 4.8 shows $f_I(x;r)$ for $r = 125\,\text{m}$ (narrow solid curve) and $r = 250\,\text{m}$ (wide solid curve), referring to middle of cell and cell-edge, respectively. The dots denote results as obtained with the approach in [92], which requires numerical evaluation of a line-integral and confirms the accuracy of the proposed finite-sum representation.

4.4. Interference and Rate at Eccentric User Locations

In (4.14), each sum term refers to a pair of transmitters $\{\mathcal{T}_{1,l}, \mathcal{T}_{1,10-l+1}\}$. The contribution of each pair to the final PDF is rendered visible by truncating the sum in (4.14) at L' with $L' \in \{1,\ldots,5\}$, i.e., only the first L' sum terms are taken into account. Dashed curves in Figure 4.8 depict results for $L' = 1$ and $L' = 2$.

It is observed that (i) in the *middle of the cell*, body and tail of the PDF are mainly shaped by the *four closest* interferers while (ii) at *cell-edge* the distribution is largely dominated by the *two closest* interferers, and (iii) interference at $r = 250$ m yields a larger variance than at $r = 125$ m due to higher diversity of the transmitter-to-user distances. The results verify link-level simulations in [174]. They emphasize the strong impact of interference asymmetry due to an eccentric user location, which is commonly overlooked in stochastic geometry analysis. In the next section, the above findings are exploited for BS coordination and -cooperation and the resulting SIR- and rate statistics are investigated.

4.4.3. Transmitter Collaboration Schemes

In thi section, SIR- and rate statistics in the full two-circle scenario, as shown in Figure 4.7, are studied. Motivated by the observations in Section 4.4.2, three schemes of *BS collaboration* are discussed:

1. *No collaboration among nodes*: This scenario represents the *baseline*, where $\mathcal{S} = \{\mathcal{T}_{0,0}\}$ and $\mathcal{I}$ comprises all nodes on the circle, i.e., $\mathcal{I} = \{\mathcal{T}_{c,n}\}$ with $c \in \{1,2\}$ and $n \in \{1,\ldots,10\}$.

2. *Interference coordination*[7]: The nodes coordinate such that co-channel interference from the two strongest interferers of the inner circle, $\mathcal{T}_{1,1}$ and $\mathcal{T}_{1,10}$, is eliminated. This could be achieved, e.g., by joint scheduling. Then, $\mathcal{S} = \{\mathcal{T}_{0,0}\}$ and $\mathcal{I}$ is composed of $\{\mathcal{T}_{1,n}\}$ with $n \in \{2,\ldots,9\}$ and $\{\mathcal{T}_{2,n}\}$ with $n \in \{1,\ldots,10\}$.

3. *Transmitter cooperation*[8]: The signals from the two closest nodes of the inner circle, $\mathcal{T}_{1,1}$ and $\mathcal{T}_{1,10}$, can be exploited as useful signals and are incoherently combined with the signal from $\mathcal{T}_{0,0}$. Then, $\mathcal{S} = \{\mathcal{T}_{0,0}, \mathcal{T}_{1,1}, \mathcal{T}_{1,10}\}$ and, as above, $\mathcal{I}$ comprises $\{\mathcal{T}_{1,n}\}$ with $n \in \{2,\ldots,9\}$ and $\{\mathcal{T}_{2,n}\}$ with $n \in \{1,\ldots,10\}$.

For each *collaboration* scheme, the PDFs of aggregate signal and -interference, $f_S(x;r)$ and $f_I(x;r)$, are calculated using Theorem 4.1. The *SIR* at user location $(r,0)$ is defined as

---
[7] Conf., e.g., Enhanced Intercell Interference Coordination (eICIC) in the 3GPP LTE-A standard [175].
[8] Conf., e.g., Coordinated Multi-Point (CoMP) in the 3GPP LTE-A standard [176].

$\gamma(r) = S(r)/I(r)$. According to [177], the PDF of $\gamma(r)$ is calculated as

$$f_\gamma(\gamma;r) = \int_0^\infty z f_S(z\gamma;r) f_I(z;r) dz, \qquad (4.20)$$

where z is an auxiliary variable, $f_S(\cdot;r)$ and $f_I(\cdot;r)$ refer to (4.9) and (4.10), and the integration bounds are obtained by exploiting the fact that $f_S(\gamma;r) = 0$ and $f_I(\gamma;r) = 0$ for $x < 0$, respectively.

Evaluating (4.9) and (4.10) yields sums of elementary functions of the form $a\gamma^b e^{-c\gamma}$, with the auxiliary parameters $a \in \mathbb{R}$, $b \in \mathbb{N}^+$ and $c > 0$. Therefore, $f_S(\gamma;r)$ and $f_I(\gamma;r)$ can generically be written as

$$f_S(\gamma;r) = \sum_s a_s \gamma^{b_s} e^{-c_s \gamma}, \qquad (4.21)$$

$$f_I(\gamma;r) = \sum_i a_i \gamma^{b_i} e^{-c_i \gamma}, \qquad (4.22)$$

and allow to straightforwardly evaluate (4.20) as

$$f_\gamma(\gamma;r) = \sum_s \sum_i \int_0^\infty z\, a_s (z\gamma)^{b_s} e^{-c_s(\gamma z)} a_i z^{b_i} e^{-c_i z} dz$$
$$= \sum_s \sum_i a_s a_i \gamma^{b_s} (c_i + c_s \gamma)^{-i-b_s-b_i} \Gamma(i + b_s + b_i). \qquad (4.23)$$

The *normalized ergodic rate* τ as a function of the SIR $\gamma(r)$ is calculated by the modified Shannon capacity formula $\tau(\gamma(r)) = \alpha_\mathrm{B} \log_2(1 + \alpha_\mathrm{SIR} \gamma(r))$, where α_B and α_SIR are coefficients for the calibration against link level simulations with $0 < \alpha_\mathrm{B} \le 1$ and $0 < \alpha_\mathrm{SIR} \le 1$, as later applied in Chapter 6. Since $\tau(\cdot)$ is a function of the RV $\gamma(r)$, its distribution is obtained by a transformation as

$$f_\tau(x;r) = \frac{1}{\alpha_\mathrm{B} \alpha_\mathrm{SIR}} 2^{x/\alpha_\mathrm{B}} f_\gamma\left(\frac{1}{\alpha_\mathrm{SIR}}\left(2^{x/\alpha_\mathrm{B}} - 1\right);r\right) \log_e(2), \qquad (4.24)$$

with $f_\gamma(\cdot;\cdot)$ from (4.23).

The distributions $f_\gamma(\gamma;r)$ and $f_\tau(\tau;r)$ are analyzed at $r = 125\,\mathrm{m}$ and $r = 250\,\mathrm{m}$ referring to *middle of the cell*, and *cell-edge*, respectively. In this chapter, $\alpha_\mathrm{B} = 1$ and $\alpha_\mathrm{SIR} = 1$. For reasons of clarity, CDF curves are presented. In order to verify the analysis, Monte Carlo simulations are carried out, employing the system model from Section 4.4.1 and the signal propagation model from Table 4.2. The results are computed by averaging over 10^7 channel realizations for each

4.4. Interference and Rate at Eccentric User Locations

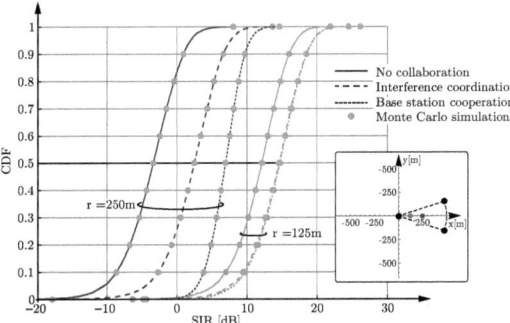

Figure 4.9.: SIR CDF curves for user locations in the *middle of the cell* ($r = 125$ m) and at *cell-edge* ($r = 250$ m), respectively. Three cases are depicted: (i) no collaboration among BSs (solid), (ii) interference coordination (dashed), (iii) cooperation among BSs (dotted).

BS collaboration scheme and each user location, and are denoted as bold dots in Figures 4.9 and 4.10, respectively.

Figure 4.9 shows the obtained SIR distributions. It is observed that

- In the case of *no collaboration* (solid lines in Figure 4.9), the curves have almost equal shape in the middle of the cell and at cell-edge. The distribution in the middle of the cell is slightly steeper due to the lower variance of the interferer impact. Their medians, hereafter used to represent the distributions' position, differ by 15.5 dB.

- When the central node $\mathcal{T}_{0,0}$ *coordinates* its channel access with the user's two dominant interferers, $\mathcal{T}_{1,1}$ and $\mathcal{T}_{1,10}$, the SIR improves by 2.4 dB in the middle of the cell and 5.9 dB at cell-edge (dashed curves in Figure 4.9), compared to *no collaboration*.

- *BS cooperation* enhances the SIR by 10.2 dB at cell-edge in comparison to *no collaboration* (left dotted curve in Figure 4.9). Note that the CDF curve also has a steeper slope than without coordination, indicating lower variance of the SIR.

- In the middle of the cell, *cooperation* achieves hardly any additional improvement, as recognized from the overlapping rightmost curves in Figure 4.9. This remarkable result states that *interference coordination* already performs close to optimal at this user location. Note that in realistic networks *coordination* is typically far less complex than *cooperation*.

Chapter 4. Modeling Asymmetric Aggregate Interference by Symmetric Structures

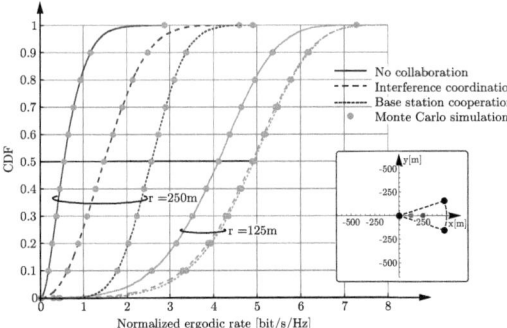

Figure 4.10.: Normalized-rate CDF curves for user locations in the *middle of the cell* ($r = 125$ m) and at *cell-edge* ($r = 250$ m), respectively. Three cases are depicted: (i) no collaboration among BSs (solid), (ii) interference coordination (dashed), (iii) cooperation among BSs (dotted).

The curves reflect findings from [178], stating that even in the best case, gains of transmitter cooperation are much smaller than largely envisioned. Figure 4.10 depicts the corresponding rate distributions. The results show that

- Notably, the rate statistics of all three collaboration schemes indicate lower variance at cell-edge than in the middle of the cell.

- In terms of median value, *BS coordination* shows rate improvements by 18.7% in the middle of the cell and by 167% at cell-edge.

- *Cooperation* between the central node $\mathcal{T}_{0,0}$ and the user's two closest interferers, $\mathcal{T}_{1,1}$ and $\mathcal{T}_{1,10}$, achieves a rate enhancement of 19.8% in the middle of the cell and 355.7% at cell-edge. Similar to the SIR, it is observed that in the middle of the cell, interference coordination already performs close to optimal.

In summary, collaboration among the BSs that were identified as main contributors to the shape of the interference distribution by Theorem 4.1, achieved large performance enhancements in terms of SIR and rate. It was further shown that the efficiency of such schemes considerably depends on the user eccentricity, or equivalently, the asymmetry of the interference impact.

4.5. Summary

In this chapter, an extended circular interference model is introduced that enables to represent substantially large interferer deployments by a well-defined circular structure in terms of interference statistics. The model applies angle-dependent power profiles along the circles, which requires the specification of a mapping procedure. A heuristic scheme is presented. Despite not claimed to be optimal, it achieves to accurately capture heterogeneous interferer deployments with hundreds or even thousands of BSs by circular models with several tens of nodes, thus reducing complexity substantially. It is observed that the accuracy of the model can be improved by increasing the number of circles rather than employing more nodes per circle.

Motivated by the desire to decompose the aggregate interference distribution into the contributions from the individual sources, a new representation for the sum of Gamma random variables with integer shape parameter is presented. The approach enables to identify candidate BSs for user-centric BS collaboration schemes and to predict the corresponding SIR- and rate statistics at eccentric user locations. Both BS-coordination and -cooperation achieve considerable performance gains in comparison to a non-collaborative scenario. It is shown that these gains largely depend on the asymmetry of the interference impact, which is either induced by non-circularly symmetric power profiles along the circles, or an arbitrary user location outside the center of the scenario.

Self-critically considering the contributions of this chapter, the following issues may be worth rethinking. Similar to the previous chapter, the framework is based on the assumption of Gamma fading. However, large-scale fading is typically modeled by a log-normal RV. Disadvantageously, its sum distribution is only accessible via estimation [107]. Hence, it would be interesting to analyze whether the assumption of Gamma fading or the error from assessing the sum of log-normal RVs leads to a higher deviation from the actual interference distribution. The accuracy of the Gamma approximation is investigated in Section 6.2.2. Considering MRT and non-coherent power accumulation for both signal and interference has been shown to overestimate the SIR [84]. A calibration against link level simulations is demonstrated in Section 6.2.3.

The presented mapping scheme provides principal directions for superior approaches, which might be reported in future work. As a first step, the Kolmogorov-Smirnov distance should be supported by additional metrics since, by itself, it does not allow for systematic optimization. The scheme is shown to improve with the number of circles C rather than the number of nodes per circle, N_c. This chapter provides guidelines for specifying C and N_c in terms of absolute numbers. It has to be scrutinized how to associate these parameters with the parameters of the scenario.

Chapter 4. Modeling Asymmetric Aggregate Interference by Symmetric Structures

In the next chapter, an approximation to investigate eccentric user locations in a stochastic system model is introduced. The particular focus lies on characterizing indoor users in dense urban environments.

Chapter 5.

Analysis of Urban Two-Tier Heterogeneous Cellular Networks

This chapter presents a system model that is based on techniques from stochastic geometry and enables the analysis of *indoor* downlink performance in urban two-tier heterogeneous cellular networks. Chapters 3 and 4 mainly deal with eccentric receiver locations, which, in general, strongly restrict stochastic models from yielding convenient expressions. In this chapter, the issue is resolved by proposing a *virtual building approximation*. Moreover, two other important limitations are addressed, namely *shadowing* and the *separation between indoor- and outdoor environments*.

In the analysis on stochastic geometry, *shadowing* is typically incorporated by log-normally distributed RVs [125, 127, 179] or neglected at all [21, 74, 119, 120, 180–184]. A recent study on blockage effects in urban environments indicates its dependency on the link length [73]. It follows the intuition that a longer link increases the likelihood of buildings to intersect with it. Such propagation characteristics have been discussed by the 3GPP only recently in a technical report on 3-dimensional channel modeling [185]. Secondly, scenarios comprising *both indoor- and outdoor environments* have not received much attention in analytical studies due to the imposed inhomogeneities on signal propagation. The designated area of operation for small cell BSs is indoors. Existing approaches either neglect the wall partitioning [122, 127], as indicated in Figure 5.1(a), oversimplify the macro-tier topology [182–184] or omit cross-tier interference [125].

In this chapter, a two-tier cellular network with outdoor macro- and indoor-deployed small cell BSs is considered. Referring to my work in [78], the contributions are:

- A tractable model for urban environment topologies is introduced. It comprises an outdoor environment, which is partly covered by circular building objects with a certain density. A method to extract its parameters from real-world data is provided. Based on concepts

Chapter 5. Analysis of Urban Two-Tier Heterogeneous Cellular Networks

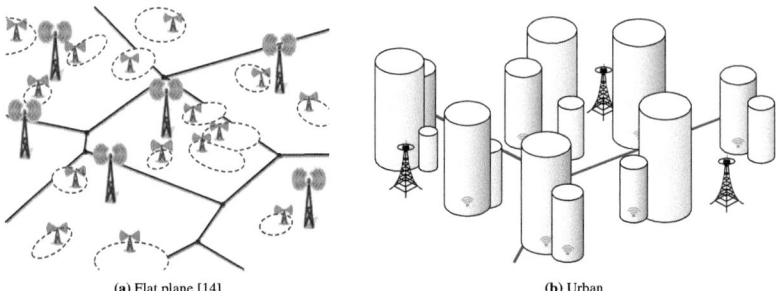

(a) Flat plane [14] (b) Urban

Figure 5.1.: Model environments of two-tier heterogeneous cellular networks.

from random shape theory, the model is applied to characterize both *signal propagation* and *network deployment*.

- A novel *virtual building approximation* to simplify aggregate interference analysis is presented. The key idea is to establish a user-centric interference environment by shifting the centers of the typical building and its exclusion regions to the user location.

- Assuming that a building is served by a small cell BS with a certain *occupation probability*, the normalized ergodic rate of a *typical* indoor user is evaluated with respect to building density and wall penetration loss. Based on these results, the impact of distinguishing Line of Sight (LOS)- and Non Line of Sight (NLOS) macro BSs is identified.

5.1. Preliminaries

5.1.1. Random Shape Theory

In this section, concepts from random shape theory are reviewed (see, e.g., [186, 187]), so as to make the model formulation in this chapter more accessible.

Let $\mathcal{O}$ denote a set of objects on $\mathbb{R}^n$, which are closed and bounded, i.e., have finite area and perimeter. For instance, $\mathcal{O}$ could be a collection of circles or rectangles on $\mathbb{R}^2$, or a combination of cubes in $\mathbb{R}^3$. For each object in $\mathcal{O}$, a *center point* is determined, which has to be well-defined but does not necessarily relate to the object's center of gravity. Non-symmetric objects additionally require to specify the orientation in space by a directional unit vector.

5.1. Preliminaries

A *Random Object Process (ROP)* is constructed by randomly sampling objects from $\mathcal{O}$ and placing their corresponding center points at the points of some PP. The orientation of each object is independently determined according to some probability distribution.

In general, a ROP is difficult to analyze, particularly when there are correlations between sampling, location and orientation of the objects. For the sake of tractability, this chapter employs a *Boolean scheme*, which satisfies the following properties: (i) the center points form a PPP, (ii) the attributes of the objects such as orientation, shape and size are mutually independent, and (iii) for each object, sampling, location and orientation are also independent.

The scheme is used to model an urban environment, where the objects of the process refer to buildings. In the next section, a method to parameterize this model with real-world data is presented.

5.1.2. Indoor Coverage Ratio

Define the *indoor coverage ratio* as the fraction of the total area on $\mathbb{R}^2$ that is covered by buildings. Then, consider a Boolean model with the center points being distributed according to a stationary PPP on $\mathbb{R}^2$ with intensity λ_B, i.e., $\Lambda(dx) = \lambda_B dx$ in (2.4). Further, let $\mathcal{C}$ denote a *random closed set* on $\mathbb{R}^2$, representing a *generic building*. According to [26, Definition 3.1.8], the resulting indoor coverage ratio is determined as

$$p_I = 1 - e^{-\lambda_B \, \mathbb{E}[|\mathcal{C}|]}, \tag{5.1}$$

where $|\cdot|$ denotes the Lebesgue measure on $\mathbb{R}^2$.

As an example, assume that $\mathcal{C}$ is a random closed ball that is centered at the origin and has random radius $R \in \mathbb{R}^+$, i.e., $\mathcal{C} = \mathcal{B}(0, R)$. Then,

$$p_I = 1 - e^{-\lambda_B \, \mathbb{E}[R^2]\pi}. \tag{5.2}$$

Expediently, the indoor coverage ratio can also be extracted from real-world data. For example, the OpenStreetMap project provides open access to shape files, as illustrated in Figure 5.2. In order to determine the degree of coverage by buildings, these files can be processed by a simple MATLAB script. For the University of Texas at Austin and Vienna's inner district (see Figures 5.2(a) and 5.2(b)) ratios of $p_I = 0.25$ and $p_I = 0.6$ were measured. Related work in [188] evaluated the indoor coverage of various Turkish cities. The ratios ranged from $0.13 - 0.39$.

In the following, the Boolean model is applied to model the deployment of a heterogeneous cellular network in accordance with the characteristics of an urban environment.

Chapter 5. Analysis of Urban Two-Tier Heterogeneous Cellular Networks

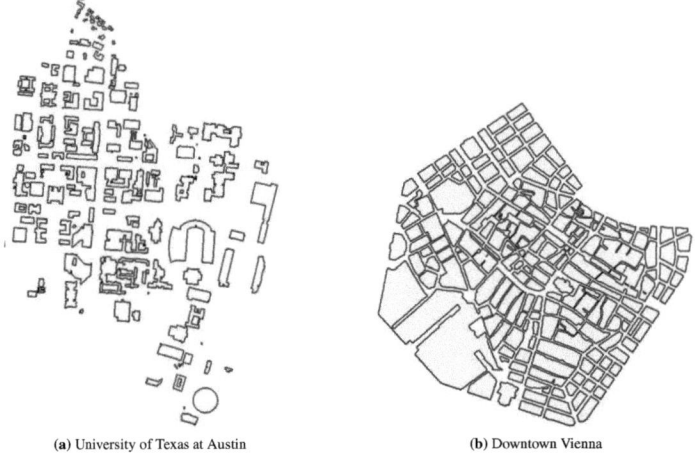

(a) University of Texas at Austin   (b) Downtown Vienna

Figure 5.2.: Building footprints as extracted from OpenStreetMap data.

5.2. System Model

5.2.1. Topology Model for Urban Environments

Consider a two-tier cellular network comprising outdoor macro- and indoor small cell BSs, as shown in Figure 5.3. Buildings are modeled by a Boolean scheme of circles on the $\mathbb{R}^2$ plane. The centers of the circles form a PPP Φ_B of intensity λ_B [26]. For simplicity, it is assumed that all circles have a fixed radius R_I. A point on the plane is said to be *indoors*, if it is covered by a building, and *outdoors* otherwise. Indoor- and outdoor environment are partitioned by a wall penetration loss, which is hereafter assumed constant for all buildings and denoted as L_W unless specified otherwise.

5.2.2. Network Deployment

Macro BSs are distributed according to a PPP Φ_M of intensity μ_M. Note that these BSs are required to be located outdoors. Thus, the macro BS process can equivalently be constructed by independently thinning[1] an initial PPP of density $\mu'_M = \mu_M/p_O$, where p_O equals the probability

---
[1] See, e.g., [26]

5.2. System Model

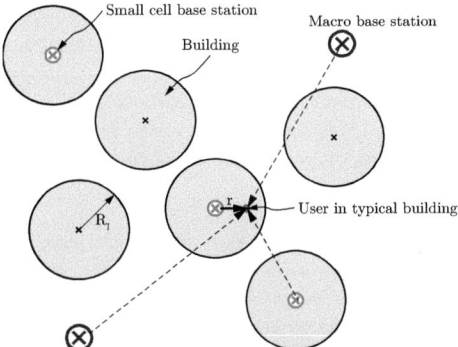

Figure 5.3.: Urban two-tier heterogeneous cellular network. Macro BSs are deployed in an outdoor environment. Buildings are modelled as a random process of circles and are assumed to have a fixed radius R_I. Only a fraction of buildings is occupied by small cell BSs. The figure depicts a typical indoor user with macro BSs and neighboring small cell BSs (dashed lines).

that a point on $\mathbb{R}^2$ is not covered by a building. According to Section 5.1.2, the thinning probability is determined as $p_\text{O} = 1 - p_\text{I} = \exp(-\lambda_\text{B} R_\text{I}^2 \pi)$.

A building will deploy an indoor small cell BS with a certain *occupation probability* η. Assume the indoor small cell BSs to be located at the center points of the occupied buildings. Then, their spatial distribution can be modeled by a PPP Φ_S of intensity $\lambda_\text{B} \eta$, which results from independently thinning the object center PPP Φ_B.

5.2.3. User Association

The aim of this chapter is to characterize the coverage and rate performance of indoor users. Noting that the buildings are assumed to form a Boolean scheme, the centers of the buildings form a PPP on the plane [73]. Therefore, by Slivnyak's theorem [26], when fixing a typical building at the origin, the centers of the other buildings still form a PPP. The performance of users will be investigated inside the typical building. Separate association rules are defined, depending on whether or not this building is occupied by a small cell BS.

Case 1 *[Typical Building with Small Cell BS]:* Consider a typical building at the origin, which is equipped with an indoor small cell BS. For simplicity, it is assumed that all users inside this building are associated with the small cell at the origin. The cases in which indoor users at the edge of the typical building may receive stronger signals from a close-by outdoor macro

BS are omitted, thus underestimating the coverage probability. Similar to the analysis in [17], *exclusion guard regions* are imposed on both macro- and small cell tier, where no BSs from the corresponding tier are allowed to distribute. For simplicity, it is assumed that the exclusion region for macro BSs is a ball of radius R_I centered at the origin, ensuring that no macro BSs are located inside the typical building. The exclusion region of the small cell tier is defined as a ball of radius $2R_I$ in order to prevent overlapping association regions of two small cells.

Case 2 *[Typical Building without Small Cell BS]:* When the typical building is not occupied by a small cell BS, the user is either associated to the dominant macro BS or a small cell BS in the immediate vicinity. The former is regarded as being of greater relevance and the latter is omitted, which leads to a lower bound on coverage probability. In this case, the indoor user will be served by the nearest BS of the macro-tier. The same exclusion regions as defined in Case 1 are employed for macro- and small cell BSs.

5.2.4. Virtual Building Approximation

Without loss of generality, a *typical indoor user* is assumed to be located at $(r, 0)$. Note that the exclusion regions as defined in Section 5.2.3 are centered at the origin rather than at the user. Consequently, the interference field as observed by the user is asymmetric and renders analysis difficult in general. Thus, the following approximation is proposed.

Let (R, θ) denote the position of an interference. Its distance to a user located at $(r, 0)$ is determined as
$$d(r) = \sqrt{R^2 + r^2 - 2Rr\cos\theta}. \tag{5.3}$$
Since typically $R \gg r$, $d(r)$ is approximated as
$$d(r) \approx R, \tag{5.4}$$
which is independent of the angle θ. As shown in Figure 5.4, the approximation in (5.4) is equivalent to shifting all the BSs along with the exclusion regions by a vector $(r, 0)$, as if the typical building was centered at the user location. Thus, this approach is referred to as *virtual building approximation*, and is applied to simplify further analysis.

5.2. System Model

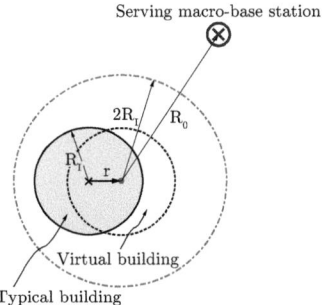

Figure 5.4.: Target area without small cell BS (gray shaded) and user-centric *virtual building* (dashed). Dashed-dotted circles denotes the shifted small cell exclusion region. The indoor user is assumed to be served by the nearest macro BS at distance R_0.

5.2.5. Signal Propagation

Macro BS to Indoor User

A signal originating from a macro BS experiences small scale fading, log-distance dependent path loss, attenuation due to building blockage and wall penetration, L_W. Small scale fading is modeled by a Gamma RV G_i, with $G_i \sim \Gamma[1,1]$, i.e., Rayleigh fading. Along the lines of [73, Theorem 1], the number of obstructing blockages along a link of length R is a Poisson RV with parameter $\beta_\text{B} R$, where $\beta_\text{B} = 2\lambda_\text{B} R_\text{I}$ in the introduced topology model. For analytical tractability the *expected blockage attenuation* as referred from [73, Theorem 6] is employed. Combining blockage- and log-distance path loss along a link of length R yields

$$\ell(R) = e^{-\beta_\text{B} R(1-L_\text{B})} \ell_\text{O}(R), \qquad (5.5)$$

where L_B refers to the attenuation of a single blockage, also denoted as *building penetration loss* and $\ell_\text{O}(R) = \min(b_\text{O}, 1/c_\text{O}\, R^{-\alpha_\text{O}})$, with intercept b_O, constant c_O and outdoor path loss exponent α_O. In this chapter, $b_\text{O} = 1$ and $c_\text{O} = 1$ for simplicity. Note that the exponential term in (5.5) incorporates the condition that the macro BS is deployed outdoors.

Eq. (5.5) reveals a major strength of the model: *Shadowing* is entirely characterized by the parameters of the underlying environment topology, which can, e.g., be extracted from openly-available online data, as demonstrated in Section 5.1.2. For comparison, the variance of log-normally distributed shadowing is typically obtained from measurements, which substantially exacerbates the finding of topologies with similar characteristics.

Small Cell BS to Indoor User

When user and small cell BS are situated in the same building, the signal experiences small scale fading and path loss $\ell_I(r) = \min(b_I, 1/c_I\, r^{-\alpha_I})$, with intercept b_I, constant c_I and indoor path loss exponent α_I. For simplicity, $b_I = 1$ and $c_I = 1$ in this chapter.

The signals from all other small cell BSs are subject to small scale fading, log-distance path loss $\ell_O(\cdot)$ as defined above, and attenuation by a factor L_W^2, as caused by the indoor-to-outdoor and outdoor-to-indoor wall penetration. Since the transmit power of a small cell BS is typically low, only small cell interferers from *neighboring* buildings are taken into account. Two buildings are defined as being *neighbors* to each other, if the segment connecting their centers is not intersected by any other building.

5.3. Performance Analysis

In this section, analytical expressions for the coverage probability of an indoor user at position $(r,0)$ are derived, regarding both buildings with- and without small cell deployment. The network is assumed to be interference limited, as is typically the case in urban areas [189]. Thus, thermal noise is neglected in the analysis.

5.3.1. Typical Building with Small Cell BS

Let $\Phi_M = \{X_i\}$ and $\Phi_S = \{X_j\}$ denote the point processes of macro- and small cell BSs, respectively. Further, define $R_i = |OX_i|$ and $R_j = |OX_j|$ as the distances of macro BS i and small cell BS j to the origin O. Assume the typical building to be occupied by a small cell BS. Then, the SIR at distance r, $0 < r \leq R_I$, is determined as

$$\gamma_S(r) = \frac{P_S G_0 \ell_I(r)}{\sum_{i:R_i \in \Phi_M \setminus \mathcal{B}(0,R_I)} P_M G_i L_W \ell(R_i) + \sum_{j:X_j \in \Phi_S \setminus \mathcal{B}(0,2R_I)} S_j P_S G_j L_W^2 \ell_O(R_j)} \quad (5.6)$$

where the terms P_M and P_S denote macro- and small cell BS transmit powers, $\ell_I(\cdot)$ and $\ell_O(\cdot)$ are indoor- and outdoor path loss laws as specified in Section 5.2.5, $\ell(\cdot)$ corresponds to the combined blockage- and path loss attenuation, as defined in (5.5), and $\mathcal{B}(0,R)$ refers to a ball of radius R, which is centered at the origin. The RVs S_j are Bernoulli distributed and, by [73, Theorem 1], have parameters $\exp(-\beta_B R_j - p_B)$, where $p_B = \lambda_B R_I^2 \pi$. They indicate whether or not an interfering small cell BS is in a neighboring building of the typical user.

Theorem 5.1. *Consider a user at distance r, $0 < r \leq R_\mathrm{I}$, away from the center of a small cell BS-occupied building. Then, its coverage probability is determined as*

$$P_{c,S}(\delta|r) = \mathbb{P}\left[\gamma_S(r) > \delta|r\right] = e^{-2\pi(\mu_S I_S + \mu_M I_M)}, \tag{5.7}$$

where

$$I_S = \int_{2R_\mathrm{I}}^{\infty} \left(\frac{\delta L_\mathrm{W}^2 \, \ell_\mathrm{O}(t) e^{-(\beta_\mathrm{B} t + p_\mathrm{B})}}{\ell_\mathrm{I}(r) + \delta L_\mathrm{W}^2 \, \ell_\mathrm{O}(t)} \right) t \, dt, \tag{5.8}$$

$$I_M = \int_{R_\mathrm{I}}^{\infty} \left(1 - \frac{\frac{P_\mathrm{S}}{P_\mathrm{M}} \ell_\mathrm{I}(r)}{\frac{P_\mathrm{S}}{P_\mathrm{M}} \ell_\mathrm{I}(r) + \delta L_\mathrm{W} \ell(t)} \right) t \, dt. \tag{5.9}$$

Proof. The proof is provided in Appendix E. □

Note that (5.8) and (5.9) correspond to the two interference contributions in (5.6).

5.3.2. Typical Building without Small Cell BS

Assume a dominant macro BS to be located at distance R_0, with $R_0 > R_\mathrm{I}$, away from the center of the typical building and consider that this building is not occupied by a small cell BS. Then, the SIR at distance r, $0 < r \leq R_\mathrm{I}$, calculates as

$$\gamma_M(R_0) = \frac{P_\mathrm{M} G_0 \ell(R_0)}{\sum_{i: R_i \in \Phi_\mathrm{M} \backslash \mathcal{B}(0, R_0)} P_\mathrm{M} G_i \ell(R_i) + \sum_{j: X_j \in \Phi_\mathrm{S} \backslash \mathcal{B}(0, 2R_\mathrm{I})} S_j P_\mathrm{S} G_j L_\mathrm{W} \ell_\mathrm{O}(R_j)} \tag{5.10}$$

Note that (i) the expression is independent of r and (ii) the factor L_W is omitted, since attenuation due to wall penetration is experienced by all signals and therefore cancels out in the SIR term.

Chapter 5. Analysis of Urban Two-Tier Heterogeneous Cellular Networks

Theorem 5.2. *Consider a user at distance r, $0 < r \le R_\mathrm{I}$, away from the center of a typical building without small cell BS and assume that it is associated with its dominant macro BS. Then, its coverage probability is determined as*

$$P_{c,\mathrm{M}}(\delta) = \mathbb{P}\left[\mathbb{E}_{R_0}\left[\gamma_\mathrm{M}(R_0) > \delta\right]\right] = \int_{R_\mathrm{I}}^{\infty} P_{c,\mathrm{M}}(\delta|R) f_{R_0}(R) dR, \quad (5.11)$$

where

$$P_{c,\mathrm{M}}(\delta|R_0) = e^{-2\pi(\mu_\mathrm{S} I_\mathrm{S} + \mu_\mathrm{M} I_\mathrm{M})}, \quad (5.12)$$

with

$$I_\mathrm{S} = \int_{2R_\mathrm{I}}^{\infty} \left(\frac{\delta L_\mathrm{W} \frac{P_\mathrm{S}}{P_\mathrm{M}} \ell_\mathrm{O}(t) e^{-(\beta_B t + p_B)}}{\ell(R_0) + \delta L_\mathrm{W} \frac{P_\mathrm{S}}{P_\mathrm{M}} \ell_\mathrm{O}(t)} \right) t\, dt, \quad (5.13)$$

$$I_\mathrm{M} = \int_{R_0}^{\infty} \left(1 - \frac{\ell(R_0)}{\ell(R_0) + \delta\, \ell(t)} \right) t\, dt, \quad (5.14)$$

and

$$f_{R_0}(R) = \begin{cases} 2\pi\mu_\mathrm{M} R\, e^{-\pi\mu_\mathrm{M}(R^2 - R_\mathrm{I}^2)} &, R \ge R_\mathrm{I} \\ 0 &, \text{otherwise} \end{cases}. \quad (5.15)$$

Proof. The conditional coverage probability $P_{c,\mathrm{M}}(\delta|R)$ in (5.12) is derived along the lines of (5.7). Averaging over the dominant macro BS distance leads to (5.11). According to [26, Example 1.4.7], the term $f_{R_0}(R)$ in (5.15) is the nearest neighbor distance distribution of a homogeneous PPP outside a ball of radius R_I. □

5.3.3. Typical Indoor User

The coverage probability of a *typical* indoor user at distance r, $0 < r \le R_\mathrm{I}$, is obtained by linearly combining $P_{c,\mathrm{S}}(\delta|r)$ from (5.7) and $P_{c,\mathrm{M}}(\delta)$ from (5.11) according to the small cell occupation probability η. Then,

$$P_c(\delta|r) = \eta\, P_{c,\mathrm{S}}(\delta|r) + (1 - \eta)\, P_{c,\mathrm{M}}(\delta). \quad (5.16)$$

Table 5.1.: Parameters for numerical evaluation.

Parameter	Value
Macro-to-small cell power ratio	$P_S/P_M = 10^{-2}$
Macro BS density	$\mu_M = 4.61 \cdot 10^{-6}$ m^{-2}
Outdoor path loss exponent	$\alpha_O = 4$
Indoor path loss exponent	$\alpha_I = 2$
Radius of building area	$R_I = 25$ m

5.4. Numerical Evaluation

In this section, the performance of a typical user at the edge of a building, i.e., $r = R_I$ is evaluated numerically. At this location, the proposed *virtual building approximation* is expected to perform worst. The normalized ergodic rate is employed as a metric. Along the lines of Section 4.4.3, it is defined as $\tau(r) = \mathbb{E}_{\gamma(r)}[\alpha_B \log_2(1 + \alpha_{SIR} \min(\delta_{max}, \gamma(r)))]$ and can be reformulated in terms of coverage probability as

$$\tau(r) = \frac{\alpha_B}{\log(2)} \int_0^{\delta_{max}} \frac{\alpha_{SIR}}{1 + \alpha_{SIR}\delta} P_c(\delta|r) d\delta, \qquad (5.17)$$

with $P_c(\delta|r)$ from (5.16) and $\delta_{max} = 2^6 - 1$, referring to 64-Quadrature Amplitude Modulation (QAM), which is the highest modulation order in the current LTE-A standard [190]. The terms α_B and α_{SIR} denote calibration parameters, with $0 < \alpha_B \le 1$ and $0 < \alpha_{SIR} \le 1$. In this chapter, $\alpha_B = 1$ and $\alpha_{SIR} = 1$.

Parameters for numerical evaluation are listed in Table 5.1. To verify the accuracy of the *virtual building approximation*, Monte Carlo simulations are carried out, using the system model as introduced in Section 5.2. The density of the macro BSs is chosen such that the inscribing ball of the typical cell has $R_C = 250$ m and the BSs are distributed over a field of $15\,R_C \times 15\,R_C$. The results are estimated from averaging over 500 fading- and 500 spatial realizations.

Figure 5.5 depicts the normalized ergodic rate $\tau(r)$ over the indoor coverage ratio p_I, as defined in Section 5.1.2. Note that when fixing the average building *size*, p_I scales with the *density* of the buildings. Solid- and dashed lines correspond to analysis and simulations, respectively. Results are shown for a sparse- and a dense small cell deployment, as quantified by the occupation probability η. For both scenarios, weak- and strong wall partitioning are investigated. The wall penetration loss is correlated to the building penetration loss L_B, as introduced in (5.5). This work employs the conservative setting $L_B = L_W$, which can be replaced by more elaborated models in future work. It is observed that

Chapter 5. Analysis of Urban Two-Tier Heterogeneous Cellular Networks

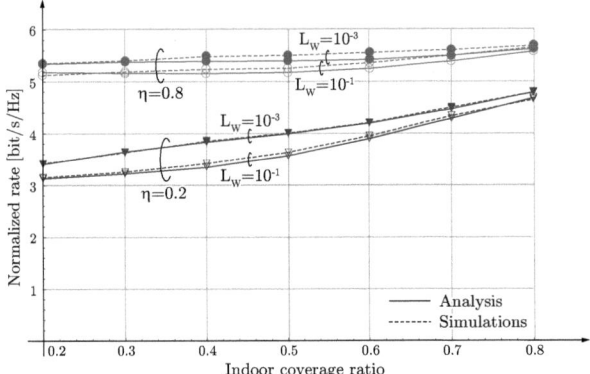

Figure 5.5.: Normalized rate [bit/s/Hz] over area-ratio, which is covered by buildings. Solid- and dashed lines denote results from analysis and simulations, respectively. Curves are shown for varying small cell occupation probability η and wall penetration loss L_W.

- The achievable normalized rate improves with increasing building density. This result follows the intuition that obstructions due to large objects establish a safeguard against interference [73]. Note that for constant occupation probability, the small cell density grows in proportion to the building density. Therefore, the results render the existence of a *hotspot limited regime* in urban environments questionable, supporting simulation results in [79, 80, 183, 184] and Chapter 6.

- Low isolation by wall penetration deteriorates performance in both deployment scenarios. Intuitively, the isolation of the interfering small cell BSs is decreased when the wall penetrations become weaker. The impact of penetration loss on coverage probability, however, becomes minor especially when the building density is high. Intuitively, this indicates that *the number of penetrations rather than the loss per penetration dominates the effect of partitioning between indoor and outdoor environment.*

- Even though a user at the edge of a typical building is evaluated, the analytical results closely resemble the simulations. This confirms the accuracy of the *virtual building approximation* as well as the inclusion of macro-interferers in the immediate vicinity of the typical building, as claimed in Section 5.2.3.

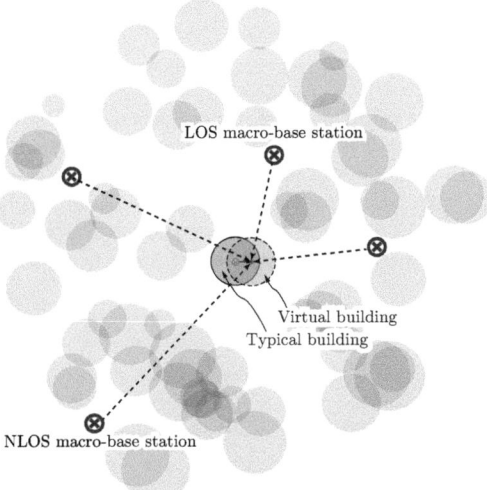

Figure 5.6.: Snapshot of an urban environment as obtained by a Boolean scheme with circles of random radius. The figure depicts an indoor user in a typical building, which is occupied by a small cell BS. The scenario encompasses LOS- and NLOS macro BSs. The dashed circle indicates the virtual building.

5.5. LOS- and NLOS Macro Base Stations

In this section, previous analyses are refined by taking into account whether the path between a macro BS and the indoor user is obstructed by any other building. The distinction between LOS- and NLOS links recently gained momentum with the study of *millimeter* wave communication [191, 192]. In contrast, it has commonly been neglected in the analysis of conventional cellular *microwave* networks. Typically, signal propagation is modeled by a single combination of shadow fading distribution, path loss law and wall penetration. Nonetheless, numerous measurement campaigns clearly indicate the differences between LOS- and NLOS microwave link characteristics [193–199].

In this section, a framework is established which enables to account for both link states. The idea is to partition the macro BSs process Φ_M into two independent non-homogeneous PPPs Φ_L and Φ_N, referring to LOS- and NLOS nodes such that $\Phi_M = \Phi_L \cup \Phi_N$. Each process employs its own model for signal propagation, incorporating log-distance dependent path loss, blockage

Chapter 5. Analysis of Urban Two-Tier Heterogeneous Cellular Networks

and wall penetration. LOS- and NLOS model are hereinafter referred to as $\ell_L(\cdot)$ and $\ell_N(\cdot)$, respectively.

Consider a typical building at the origin and an indoor user at distance r, $0 < r \leq R_I$. Further, let a macro BS be located at distance R away from the origin. Then, using the Boolean scheme as specified in Section 5.2.1 and the *virtual building approximation* as introduced in Section 5.2.4, the probability that the link between user and BS is obstructed by any other building is obtained as $v(R) = \exp(-\beta_B(R - R_I))$. This term can be interpreted as *LOS probability*, following the intuition that it becomes increasingly unlikely to experience a LOS connection with a distant BS.

According to [73], the shadowing of different links can be considered as uncorrelated with minor loss of accuracy. This allows to apply the thinning property of PPPs (conf. Section 2.1.2) and yields the intensities of the independent BS processes Φ_L and Φ_N as $\mu_M v(R)$ and $\mu_M(1-v(R))$, respectively.

5.5.1. Distance Distributions of Associated Macro Base Stations

Consider a typical building without small cell BS. The following lemmas provide PDFs for the distance between an indoor user and its associated macro BS, given that the BS is either in LOS or NLOS. The expressions extend results in [191] by conditioning on deploying a *virtual building* around the user, as illustrated in Figure 5.6.

Lemma 5.1. *Consider a typical building without a small cell BS and an indoor user at distance r, $0 < r \leq R_I$, away from its center, which is associated with the closest LOS macro BS. Then, applying the virtual building approximation, its distance to the serving BS is distributed as*

$$f_L(x) = \begin{cases} \frac{B_L \hat{f}_L(x)}{A_L} e^{-2\pi \mu_M \int_{R_I}^{\Psi_L(x)} (1-v(t))t dt} &, x \geq R_I \\ 0 &, \text{otherwise} \end{cases} \quad (5.18)$$

where

$$B_{\mathrm{L}} = 1 - e^{-2\pi\mu_{\mathrm{M}} \int_{R_{\mathrm{I}}}^{\infty} v(r) r dr} \tag{5.19}$$

$$\hat{f}_{\mathrm{L}}(x) = \begin{cases} \frac{1}{B_{\mathrm{L}}} 2\pi\mu_{\mathrm{M}} x v(x) e^{-2\pi\mu_{\mathrm{M}} \int_{R_{\mathrm{I}}}^{x} v(r) r dr} &, x \geq R_{\mathrm{I}} \\ 0 &, \text{otherwise} \end{cases} \tag{5.20}$$

$$A_{\mathrm{L}} = B_{\mathrm{L}} \int_{R_{\mathrm{I}}}^{\infty} e^{-2\pi\mu_{\mathrm{M}} \int_{R_{\mathrm{I}}}^{\Psi_{\mathrm{L}}(x)} (1-v(t)) t dt} \hat{f}_{\mathrm{L}}(x) dx, \tag{5.21}$$

and $\Psi_{\mathrm{L}}(x) = \ell_{\mathrm{L}}^{-1}(\ell_{\mathrm{N}}(x))$. The term B_{L} denotes the probability that the user receives at least one LOS BS and $\hat{f}_{\mathrm{L}}(x)$ is the corresponding conditional distance distribution function of the closest node. The quantity A_{L} captures the likelihood to be associated with the closest LOS BS.

Proof. The proof is derived along the lines of [73, Theorem 10] by excluding BS from a ball of radius R_{I} around the user. □

Lemma 5.2. *Consider an indoor user at distance r, $0 < r \leq R_{\mathrm{I}}$ away from the center of a typical building without a small cell BS. Let the user be attached to the closest NLOS BS. Then, employing the virtual building approximation, the PDF of its distance to the serving BS is expressed as*

$$f_{\mathrm{N}}(x) = \begin{cases} \frac{B_{\mathrm{N}} \hat{f}_{\mathrm{N}}(x)}{A_{\mathrm{N}}} e^{-2\pi\mu_{\mathrm{M}} \int_{R_{\mathrm{I}}}^{\Psi_{\mathrm{N}}(x)} v(t) t dt} &, x \geq R_{\mathrm{I}} \\ 0 &, \text{otherwise} \end{cases}, \tag{5.22}$$

where

$$B_{\mathrm{N}} = 1 - e^{-2\pi\mu_{\mathrm{M}} \int_{R_{\mathrm{I}}}^{\infty} (1-v(r)) r dr} \tag{5.23}$$

$$\hat{f}_{\mathrm{N}}(x) = \begin{cases} \frac{1}{B_{\mathrm{N}}} 2\pi\mu_{\mathrm{M}} x (1-v(x)) e^{-2\pi\mu_{\mathrm{M}} \int_{R_{\mathrm{I}}}^{x} (1-v(r)) r dr} &, x \geq R_{\mathrm{I}} \\ 0 &, \text{otherwise} \end{cases} \tag{5.24}$$

$$A_{\mathrm{N}} = 1 - A_{\mathrm{L}}, \tag{5.25}$$

with A_{L} from (5.21) and $\Psi_{\mathrm{N}}(x) = \ell_{\mathrm{N}}^{-1}(\ell_{\mathrm{L}}(x))$. The term B_{N} refers to the probability that the user receives at least one NLOS BS and $\hat{f}_{\mathrm{N}}$ is the according conditional PDF of the distance to the closest node.

Proof. As above, the proof follows [73, Theorem 10] and is therefore omitted. □

5.5.2. SINR and Coverage Analysis

Given an indoor user at distance r, $0 < r \leq R_\text{I}$ away from the center of a typical building with small cell BS, its SIR is determined as

$$\gamma_\text{S}(r) = \frac{P_\text{S} G_0 \ell_\text{I}(r)}{\sum\limits_{\substack{i: X_i \in \Phi_\text{L} \\ \setminus \mathcal{B}(0, R_\text{I})}} P_\text{M} G_i \ell_\text{L}(R_i) + \sum\limits_{\substack{j: X_j \in \Phi_\text{N} \\ \setminus \mathcal{B}(0, R_\text{I})}} P_\text{M} G_j \ell_\text{N}(R_j) + \sum\limits_{\substack{k: X_k \in \Phi_\text{S} \\ \setminus \mathcal{B}(0, 2R_\text{I})}} S_k P_\text{S} G_k L_\text{W} \ell_\text{L}(R_k)}, \quad (5.26)$$

where the first- and second sum in the denominator denote the aggregate interference from the LOS- and NLOS macro BSs, respectively. The third sum refers to the contribution from the small cell tier.

When the building is not occupied by a small cell BS, the user associates with the dominant macro BS at distance R_0, with $R_0 > R_\text{I}$. Depending on whether the serving BS is in LOS or NLOS, (5.10) reformulates as

$$\gamma_\text{L}(R_0) = \frac{P_\text{M} G_0 \ell_\text{L}(R_0)}{\sum\limits_{\substack{i: X_i \in \Phi_\text{L} \\ \setminus \mathcal{B}(0, R_0)}} P_\text{M} G_i \ell_\text{L}(R_i) + \sum\limits_{\substack{j: X_j \in \Phi_\text{N} \\ \setminus \mathcal{B}(0, \Psi_\text{L}(R_0))}} P_\text{M} G_j \ell_\text{N}(R_j) + \sum\limits_{\substack{k: X_k \in \Phi_\text{S} \\ \setminus \mathcal{B}(0, 2R_\text{I})}} S_k P_\text{S} G_k L_\text{W} \ell_\text{L}(R_k)},$$

$$(5.27)$$

or

$$\gamma_\text{N}(R_0) = \frac{P_\text{M} G_0 \ell_\text{N}(R_0)}{\sum\limits_{\substack{i: X_i \in \Phi_\text{L} \\ \setminus \mathcal{B}(0, \Psi_\text{N}(R_0))}} P_\text{M} G_i \ell_\text{L}(R_i) + \sum\limits_{\substack{j: X_j \in \Phi_\text{N} \\ \setminus \mathcal{B}(0, R_0)}} P_\text{M} G_j \ell_\text{N}(R_j) + \sum\limits_{\substack{k: X_k \in \Phi_\text{S} \\ \setminus \mathcal{B}(0, 2R_\text{I})}} S_k P_\text{S} G_k L_\text{W} \ell_\text{L}(R_k)}.$$

$$(5.28)$$

Note that by virtue of the virtual building approximation from Section 5.2.4 both $\gamma_\text{L}(\cdot)$ and $\gamma_\text{N}(\cdot)$ are independent of the user's location within the building. The following theorems extend Theorems 5.1 and 5.2 with respect to LOS- and NLOS macro BSs.

Theorem 5.3. *Consider an indoor user at distance r, $0 < r \leq R_\text{I}$ away from the center of a typical building with a small cell BS. Then, its coverage probability is calculated as*

$$P_{c,\text{S}}(\delta|r) = \mathbb{P}\left[\gamma_\text{S}(r) > \delta|r\right] = e^{-2\pi(\mu_\text{S} I_\text{S} + \mu_\text{M}(I_\text{L} + I_\text{N}))}, \quad (5.29)$$

with $\gamma_S(\cdot)$ from (5.26) and

$$I_S = \int_{2R_I}^{\infty} \frac{\delta L_W \ell_L(t) e^{-(\beta_B t + p_B)}}{\ell_I(r) + \delta L_W \ell_L(t)} t dt, \qquad (5.30)$$

$$I_L = \int_{R_I}^{\infty} \left(1 - \frac{\frac{P_S}{P_M}\ell_I(r)}{\frac{P_S}{P_M}\ell_I(r) + \delta \ell_L(t)}\right) tv(t) dt, \qquad (5.31)$$

$$I_N = \int_{R_I}^{\infty} \left(1 - \frac{\frac{P_S}{P_M}\ell_I(r)}{\frac{P_S}{P_M}\ell_I(r) + \delta \ell_N(t)}\right) t(1 - v(t)) dt. \qquad (5.32)$$

Proof. The proof is provided in Appendix F. □

Theorem 5.4. *Consider a typical building without a small cell BS and an indoor user at distance r, $0 < r \le R_I$ away from its center. Given that the user is associated with the closest LOS macro BS, its coverage probability is determined as*

$$P_{c,L}(\delta) = \mathbb{P}\left[\mathbb{E}_{R_0}[\gamma_L(R_0) > \delta]\right] = \int_{R_I}^{\infty} e^{-2\pi(\mu_S I_S + \mu_M(I_L + I_N))} f_L(R) dR, \qquad (5.33)$$

with $\gamma_L(\cdot)$ from (5.27), $f_L(\cdot)$ from (5.18) and

$$I_S = \int_{2R_I}^{\infty} \frac{\delta P_S L_W \ell_L(t) e^{-(\beta_B t + p_B)}}{P_M \ell_L(R) + \delta P_S L_W \ell_L(t)} t \, dt \qquad (5.34)$$

$$I_L = \int_R^{\infty} \left(1 - \frac{\ell_L(R)}{\ell_L(R) + \delta \ell_L(t)}\right) t \, v(t) dt \qquad (5.35)$$

$$I_N = \int_{\Psi_L(R)}^{\infty} \left(1 - \frac{\ell_L(R)}{\ell_L(R) + \delta \ell_N(t)}\right) t \, (1 - v(t)) dt. \qquad (5.36)$$

When the user is served by the closest NLOS macro BS, its coverage probability is calculated as

$$P_{c,N}(\delta) = \mathbb{P}\left[\mathbb{E}_{R_0}[\gamma_N(R_0) > \delta]\right] = \int_{R_I}^{\infty} e^{-2\pi(\mu_S I_S + \mu_M(I_L + I_N))} f_N(R) dR \qquad (5.37)$$

where $\gamma_N(\cdot)$ and $f_N(\cdot)$ are obtained from (5.22) and (5.28), and

$$I_S = \int_{2R_I}^{\infty} \frac{\delta P_S L_W \ell_L(t) e^{-(\beta_B t + p_B)}}{P_M \ell_N(R) + \delta P_S L_W \ell_L(t)} t\, dt \tag{5.38}$$

$$I_L = \int_{\Psi_N(R)}^{\infty} \left(1 - \frac{\ell_N(R)}{\ell_N(R) + \delta \ell_L(t)}\right) t\, v(t) dt \tag{5.39}$$

$$I_N = \int_R^{\infty} \left(1 - \frac{\ell_N(R)}{\ell_N(R) + \delta \ell_N(t)}\right) t\, (1 - v(t)) dt. \tag{5.40}$$

Proof. For a given BS distance R_0, the proofs for $\mathbb{P}[\gamma_N(R_0) > \delta]$ and $\mathbb{P}[\gamma_L(R_0) > \delta]$ are carried out along the lines of (F.5) in Appendix F. Averaging over R_0 yields (5.33) and (5.37), respectively. □

Finally, Theorems 5.3 and 5.4 enable to extend (5.16). The coverage probability of a *typical* indoor user at distance r, $0 < r \le R_I$, which experiences LOS- and NLOS macro BSs, is expressed as

$$P_c(\delta|r) = \eta P_{c,S}(\delta|r) + (1 - \eta)(A_L P_{c,L}(\delta) + (1 - A_L) P_{c,N}(\delta)), \tag{5.41}$$

where η denotes the small cell occupation probability and A_L is the likelihood that the user associates with a LOS macro BS, as derived in (5.21).

5.5.3. Numerical Evaluation

In this section, the performance of a typical user at the edge of a building, i.e., $r = R_I$, is numerically evaluated. The results are provided in terms of normalized ergodic rate and are obtained by plugging $P_c(\delta|R)$ from (5.41) into (5.17).

Signal propagation along LOS- and NLOS links is modeled by

$$\ell_L(R) = \ell_O(R) L_L, \tag{5.42}$$

$$\ell_N(R) = e^{-\beta_B R(1 - L_B)} \ell_O(R) L_N, \tag{5.43}$$

where L_L and L_N denote the wall penetration losses, $\exp(-\beta_B R(1 - L_B))$ accounts for the shadowing, as referred from Section 5.2.5, and $\ell_O(\cdot)$ refers to the log-distance path loss, as defined in Section 5.2.5. Note that, in general, the intercept b_O, the constant c_O and the path loss exponent α_O will be different in the LOS- and NLOS case. These models are based on the following findings from measurement campaigns.

For analytical convenience, the characteristics of an urban environment are often condensed into different variances of a log-normally distributed RVs, which account for the shadowing [200]. However, the authors of [193] observed considerable deviations from this model in LOS scenarios, where signal characteristics are largely dominated by free space propagation as long as the first Fresnel zone is not obstructed [193–197]. Breaking distances of 160 - 800 m have been reported from measurements in metropolitan areas [196, 201]. Using the concept of a *LOS ball* as defined in [191] yields equivalent circular LOS areas of radius 49.1 m for p_I = 0.8, 96.2 m for p_I = 0.5 and 266 m for p_I = 0.2, respectively. Hence, it is considered reasonably accurate to employ a single-slope free-space path loss law for LOS signal propagation, i.e., α_O = 2 in (5.42).

In accordance with the 3GPP LTE-A standard [202] and numerous measurement campaigns [194–197, 203, 204], NLOS propagation alters the path loss exponent and adds an additional shadowing component. In this section, α_O = 4 and L_B = 10^{-1}, respectively. Furthermore, measurement results in [193, 198, 199] indicate that signals from a NLOS BS experience a lower wall penetration loss. Intuitively, this is caused by the fact that multi path components approach the building more frontally after multiple reflections. The wall penetration loss for NLOS links is set L_N = 10^{-1} whereas the loss for LOS links, L_L, is varied as specified below. Table 5.2 summarizes the parameters for numerical evaluation.

In order to verify the accuracy of the *virtual building approximation*, Monte Carlo simulations are carried out with the same system model. BS are distributed over a field of 15 R_C × 15 R_C and their density is chosen such that the inscribing ball of the typical cell has a radius of 250 m. The results are assessed from averaging over 500 spatial- and 500 fading realizations.

Figure 5.7 depicts normalized ergodic rate $\tau(R_I)$ over indoor coverage ratio p_I. The curves correspond to two small cell occupation probabilities, η = 0.2 and η = 0.8, and two wall penetration loss values for the LOS signal, L_L = 10^{-1} and L_L = 10^{-3}, respectively.

It is observed that

- Unlike results in Section 5.4, the normalized rate does not increase uniformly with the indoor coverage ratio but rather exhibits certain minima. This is explained by the facts that (i) a low indoor coverage ratio favors LOS connections and (ii) the likelihood to experience a LOS link rapidly decreases with higher indoor coverage ratio as the exponent of the LOS probability $v(R)$ linearly scales with the building density.

Table 5.2.: Parameters for numerical evaluation.

Parameter	Value
Macro-to-small cell power ratio	$P_S/P_M = 10^{-2}$
Macro BS density	$\mu_M = 4.61 \cdot 10^{-6}$ m^{-2}
Radius of building area	$R_I = 25$ m
Intercept of path loss law	$b_O = 1$
Constant of path loss law	$c_O = 1$
LOS path loss exponent	$\alpha_O = 2$
NLOS path loss exponent	$\alpha_O = 4$
Building penetration loss	$L_B = 10^{-1}$
NLOS wall penetration loss	$L_N = 10^{-1}$

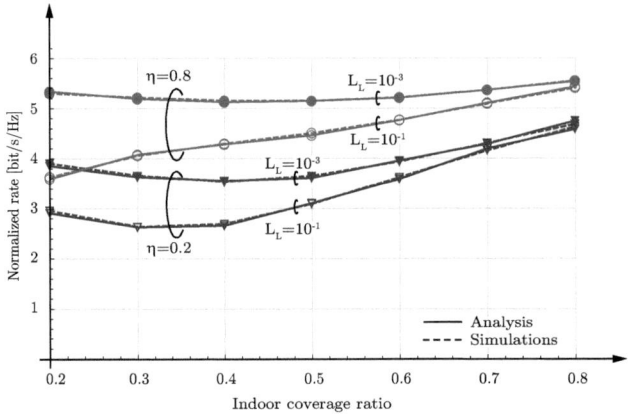

Figure 5.7.: Normalized rate in [bit/s/Hz] over indoor coverage ratio. Solid- and dashed lines correspond to results from analysis and simulations, respectively. Curves are depicted for varying small cell occupation probability η and wall penetration loss for LOS links, L_L, respectively.

- In a scenario with low wall penetration loss for LOS links, small cell BSs are weakly isolated from the outdoor environment. Hence, for a high small cell occupation probability, they considerably contribute to the aggregate interference. As a result, the normalized rate for $L_\mathrm{L} = 10^{-1}$ and $\eta = 0.8$ monotonically increases rather than exhibiting a minimum.

- The impact of the wall penetration loss becomes inferior with higher indoor area coverage. The result is based on the effects that (i) the likelihood of experiencing LOS interferers reduces with decreasing LOS probability and (ii) a higher building density establishes a safeguard against NLOS interferers. This corroborates previous findings that it is *the amount of blockages rather than their exact penetration loss which dominates the effect of indoor-outdoor partitioning*.

- The curves precisely fit with results from Monte Carlo simulations. In comparison to Section 5.4, the consideration of LOS- and NLOS macro BSs improves the accuracy of the model.

As shown in Chapter 6, these effects cannot be identified with the widely used log-normal shadowing model.

5.6. Summary

This chapter introduces a novel system model for two-tier heterogeneous cellular networks in urban environments. The focus lies on indoor users. Analytical expressions for the coverage probability in buildings with- and without small cell deployment are derived. The proposed *virtual building approximation* considerably improves the tractability of the analysis. Its accuracy is confirmed by simulation results. Numerically evaluations are carried out to investigate the performance of a *typical* indoor user in terms of normalized ergodic rate. The results reveal essential effects of an urban environment. Observations such as the blockage safeguard and the vanishing impact of LOS BSs and wall isolation with increasing building density have been missed by overly simplistic models.

In a self-critical retrospective, the following points may be worth rethinking. The major weakness of the model is the negligence of reflections, thus ignoring effects of gracing incidence and wave guidance, as reported e.g., in [205]. At best, these effects are incorporated in the path loss exponent. Their explicit treatment provides an interesting topic for future work.

The model does not account for multiple small cell BSs within the same building. In future work this might be included by allowing buildings to overlap with the typical building at the origin. Then, a small cell BS in an overlapping building could be considered as *intra-building* interferer. Furthermore, the investigations do not account for walls within buildings. It has to be further investigated whether they can be modeled by a ROP in a similar fashion as the buildings, allowing to scrutinize users that are located *deep indoors* and *close to an outer wall*, respectively.

The introduced model does not consider small cell operation modes such as Open Subscriber Group (OSG) and Closed Subscriber Group (CSG). As shown in Sections 6.3.2–6.3.4, these association policies have a considerable impact on the performance. Hence, it would be of high interest to incorporate them in the model.

Recent work in [206] exploits the *displacement theorem* to incorporate log-normally distributed shadowing in the stochastic geometry analysis. A comparison with the results as presented in this chapter would be instructive to reveal deviations from a well-established model. Section 6.3.2 provides insights by means of LTE-A system level simulations.

In this chapter, outage is defined as the event that the SIR deceeds a certain threshold. Heterogeneous networks may exhibit massive variations between nominal cell sizes, thus making the cell-load an important determinant for the achievable rate [7]. Hence, a more relevant definition for future network deployments might claim outage, when a certain minimum-rate is not achieved. Its evaluation by means of the introduced framework is an interesting extension for further work.

The chapter does not address the typical *outdoor* user which would be substantial to gain full understanding of the performance of a two-tier heterogeneous cellular network in an urban environment. The next chapter provides system level simulation studies to complete the picture.

Chapter 6.
LTE-A System Level Simulations

In this chapter, the theory from the previous three chapters is supported by system level simulations in a realistic LTE-A environment. In addition to the hitherto *user-centric* considerations, *network-wide* performance metrics are also evaluated. The simulations are carried out with the Vienna LTE-A downlink system level simulator (latest version: v1.8 r1375), which is fully compliant with the 3GPP standard.

The first part of the chapter introduces the Vienna LTE-A simulator and provides details on its physical layer modeling. In the second part, results from Chapters 3 and 4 are reproduced, using a homogeneous macro site deployment, which is arranged according to a hexagonal grid. In the third part of the chapter, the focus is on heterogeneous LTE-A networks. The goal here is to reconsider results from Chapter 5 and to gain insights on the *global* impact of deploying small cells in an existing macro cellular network. The common simulation parameters are largely adopted from [167] and listed in Table 6.1. As the chapter deals with LTE-A, the terms *eNodeB* and *User Equipment (UE)* are used instead of *BS* and *user*, respectively. In the heterogeneous deployments, small cells are represented by *femtocells*.

6.1. Vienna LTE-A Downlink System Level Simulator

Performance evaluation on system level typically encompasses a large number of network elements and upscales the number of interconnecting links [81]. Hence, computational complexity needs to be decreased substantially in order to make the problem feasible. The Vienna LTE-A downlink system level simulator employs a Mutual Information based exponential SNR Mapping (MIESM) for link abstraction [207, 208], which allows to reduce the resolution of the time-frequency grid and omits protracted encoding- and decoding procedures. Historically, the first setup based on MIESM was introduced by Josep Colom Ikuno and Martin Wrulich in 2010 [53]. Since then, the major enhancements that were carried out by me include the

Chapter 6. LTE-A System Level Simulations

Table 6.1.: Common simulation parameters.

Parameter	Value
Carrier frequency	f_c = 2.14 GHz
LTE-A bandwidth	20 MHz
Macro site deployment	hexagonal grid, one ring
Inter-macro site distance	500 m
eNodeB transmit power	P_M = 46 dBm
eNodeB minimum coupling loss	c_B = −70 dB
Shadow fading	spatially-correlated log-normal
Fast fading	time-correlated Rayleigh
Receiver type	zero forcing
Noise power density	−174 dBm/Hz
Traffic model	full buffer
Channel knowledge	perfect
Simulation length	100 Transmission Time Interval (TTI)
Number of simulation runs	100

support of heterogeneous cellular networks [79, 80], stochastic network deployments and CoMP techniques [84, 85], applying the concept of *runtime precoding* [81].

The simulator is implemented in object-oriented MATLAB[1] and is made *openly available* for download under an academic, non-commercial use license. Its rich set of features and easy adaptability has led to numerous publications from researchers all over the globe, including studies on energy-efficient cell-coordination schemes [210], handover algorithms in self-optimizing networks [211], and resource allocation techniques for femtocell networks [212] as well as for machine-to-machine communication [213]. On top of that, the *open accessibility* warrants the reproducibility of these contributions. Today (May 2015), the simulator counts more than 30 000 downloads and undergoes permanent peer-review from a substantially large online community. With some 100 000 lines of code, employing a large forum with active users is the only method to guarantee its quality. The remainder of this section briefly explains the simulator's *physical layer modeling*. For a comprehensive description, the interested reader is referred to [214].

The LTE-A PHY procedures can conceptually be described as a BICM-system [214], as shown in Figure 6.1. It comprises a transmitter including channel coder, bit interleaver and modulator ($\mathcal{M}$). In LTE-A, coding and interleaving is achieved by a turbo-coder in combination with rate matching. The symbol mapping employs 4-, 16- and 64-QAM with Gray mapping, respectively. Signal propagation over an $N_{Rx} \times N_{Tx}$ Multiple-Input-Multiple-Output (MIMO) channel is modeled by slowly-varying, position-dependent macro scale fading, small-scale fading and

---

[1] For further information see, e.g., [209].

6.1. Vienna LTE-A Downlink System Level Simulator

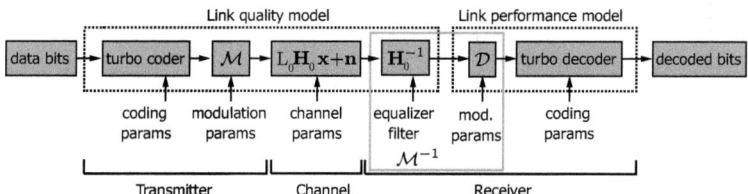

Figure 6.1.: Separation of an LTE-A link into link quality- and link performance model. The link can equivalently be described as an LTE BICM transmitter-receiver chain [214].

Additive White Gaussian Noise (AWGN). It can be represented by a complex-valued $N_{\text{Rx}} \times N_{\text{Tx}}$ matrix, which follows from the assumption that the cyclic prefix exceeds the channel length, hence omitting inter-symbol interference. The receiver encompasses an equalizer filter and a demodulator ($\mathcal{M}^{-1}$) as well as a turbo decoder, which provides de-interleaving and channel decoding. In the current version of the simulator, low complexity models for Zero Forcing (ZF)- and Minimum Mean Square Error (MMSE) receivers are available. The former approaches the average performance of an optimal receiver by exploiting Multi-User (MU) diversity, which is typically present in system level scenarios [215].

The objective of the *link abstraction model* is to predict the performance of the presented LTE-A link, given a parameterization of the inputs. For simplification, the model can be divided into a *link quality-* and a *link performance model*, as indicated in Figure 6.1. The link quality model measures the quality of the received signal after equalization. Since in the presented case, the metric has to represent the quality of the input to the turbo decoder, the post-equalization SINR is a straightforward choice [214]. The link performance model translates this measure into Block-Error Ratio (BLER) and further into (area) spectral efficiency and effective throughput, based on the employed Modulation and Coding Scheme (MCS). On top of that, the Vienna LTE-A supports Hybrid Automatic Repeat Request (HARQ) (applying a hybrid Chase combining/incremental redundancy approach) for a more realistic characterization of the link [216].

The model in Figure 6.1 is a simplification of the actual link abstraction model, as it does not account for interference from other base stations. Its expansion to the whole network, as applied in the Vienna LTE-A simulator, is illustrated in Figure 6.2. The figure identifies the main components of the model as network layout, time-variant fading and scheduling. It also illustrates the corresponding input-output relations to the link quality- and link performance model, respectively.

The Vienna LTE-A simulator employs a MIESM for the SINR-to-BLER mapping [207, 208],

Chapter 6. LTE-A System Level Simulations

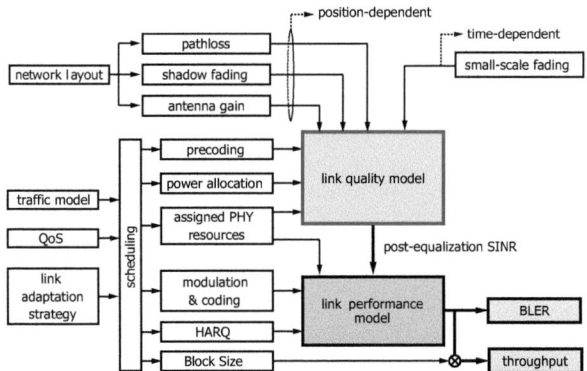

Figure 6.2.: LTE-A link abstraction model with new *link quality model* as employed in the Vienna LTE-A downlink system level simulator v1.8 r1375.

which already proved beneficial in Release 5 of UMTS [217]. This method compresses the SINR values of the assigned Resource Blocks (RBs) for each UE and 1 ms-long subframe (subsequently also denoted as TTI) into an *effective SINR*, yielding an AWGN-equivalent representation in terms of mutual information. These SINR values are then mapped to a BLER by means of an AWGN BLER curve of the corresponding MCS. The curves are obtained from LTE-A link level simulations, thus forming the *only* computationally costly physical layer evaluation, which is required for the link abstraction model.

The simulator achieves further complexity reduction by employing a block fading model, i.e., assuming a constant channel for the duration of one TTI and by representing each RB by only 2 (out of 12 possible) subcarrier post-equalization SINR values. For instance, an LTE-A bandwidth of 20 MHz yields 200 (instead of 1200) SINR samples per TTI as an output of the link quality model. In [214], accurate abstractions of LTE Transmission Modes (TMs) 1-4 were presented. Compared to link level simulations, the speed-up values in observed runtime ranged from hundredfold up to thousandfold while preserving highly consistent results. With the introduction of the *runtime-precoding* concept in [81], the list of supported TMs in the Vienna LTE-A simulator was extended towards TM 9, while keeping the additional computational expense at a minimum. In the remainder of this chapter, closed-loop spatial multiplexing (TM 4) is employed.

6.2. Homogeneous Macro Cellular Network

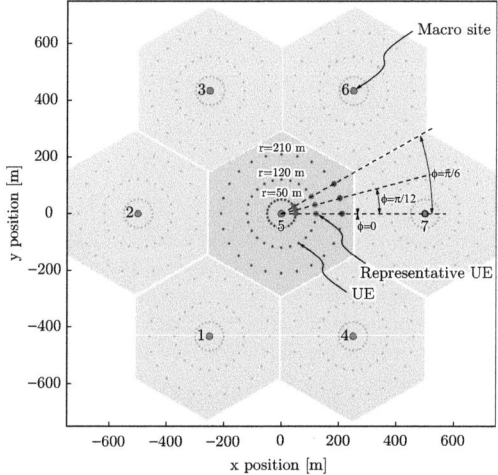

Figure 6.3.: Hexagonal grid setup with central cell and six interfering eNodeBs. UEs are equidistantly distributed along circles of radius r = {50, 120, 210} m. Bold dots indicate representative UE positions. The corresponding angles are given by $\phi = \{0, \frac{\pi}{12}, \frac{\pi}{6}\}$, respectively. In the case of BS collaboration, eNodeB 7 does not contribute to the aggregate interference.

6.2. Homogeneous Macro Cellular Network

In this section, the validity of the Gamma distribution for approximating aggregate interference in symmetric interference scenarios, as presented in Chapter 3, and the impact of asymmetric interference, as investigated in Chapter 4, is evaluated, respectively. In the first part, the corresponding system model is introduced.

6.2.1. System Model

The system model is composed of a central macro site and six neighboring nodes, which are arranged according to a hexagonal grid, as illustrated in Figure 6.3. Each site employs a single eNodeB, which is equipped with an omni-directional antenna. For systematic investigations, the UEs are equidistantly distributed along concentric circles of radius $r = \{50, 120, 210\}$ m, referring to cell-center, middle of cell and cell-edge, respectively. Each circle encompasses 24 UEs, which are uniquely identified by the tuple (r, ϕ), where ϕ denotes the angle position. The

Chapter 6. LTE-A System Level Simulations

Table 6.2.: Specific parameters for simulations of homogeneous macro cellular network.

Parameter	Value
Antenna configuration	$N_{\text{Tx}} \times N_{\text{Rx}} = 1 \times 1$
eNodeB antenna gain in dB	$A(\theta) = 0\,\text{dB}$
Path loss	$\ell(d) = \min(b_{\text{P}}, 1/c_{\text{P}}\, d^{-2})$
Scheduler type	round robin

signal experiences free-space path loss[2], fast fading according to a time-correlated Rayleigh channel, and spatially-correlated log-normal shadowing with 8 dB standard deviation[3]. Hereinafter, the combination of these three mechanisms is termed *composite fading*. The results in this section are obtained by averaging over 100 channel realizations and 100 TTIs. The simulation parameters are summarized in Tables (6.1) and (6.2), respectively.

6.2.2. Validation of Gamma Approximation

In this section, the circular interference model as presented in Chapter 3 is validated. The particular focus lies on the accuracy of the Gamma distribution as an approximation for both composite fading and aggregate interference. The system model is referred from Section 6.2.1.

Firstly, the average aggregate interference is measured along each of the three UE circles. The results are depicted as solid lines in Figure 6.4. In accordance with Section 3.4.1, it is observed that average aggregate interference is almost constant at the cell-center and in the middle of the cell. At cell-edge, the curves exhibit fluctuations due to the vicinity of the dominant interferers. Results from the circular interference model accurately assess the average behavior, as shown by the dashed lines.

In the next step, the Empirical Cumulative Distribution Function (ECDF) of the aggregate interference is computed at nine representative UE locations, which are marked by bold dots in Figure 6.3. Similar to Section 3.4.2, the angle positions $\phi = \{0, \frac{\pi}{6}, \frac{\pi}{12}\}$ refer to UEs with one dominant interferer (eNodeB 7), two equidistant dominant interferers (eNodeBs 6 and 7) and a variation thereof. Solid lines in Figure 6.5 depict the results. In accordance with Section 3.4.2, the interference distributions are dominated by the UEs' distances to the origin while their angle positions have only minor impact. The latter is illustrated by the enlarged section in Figure 6.5.

---

[2] The free-space path loss law is defined as $\min(b_{\text{P}}, 1/c_{\text{P}} d^{-2})$. In this section, $b_{\text{P}} = 10^{-7}$ and $f_c = 2.14\,\text{GHz}$, yielding $c_{\text{P}} = (4\pi f_c/c_0)^2 = 8.0465 \cdot 10^3$, where c_0 is the speed of light.

[3] The shadow fading maps are computed by applying the method in [218].

6.2. Homogeneous Macro Cellular Network

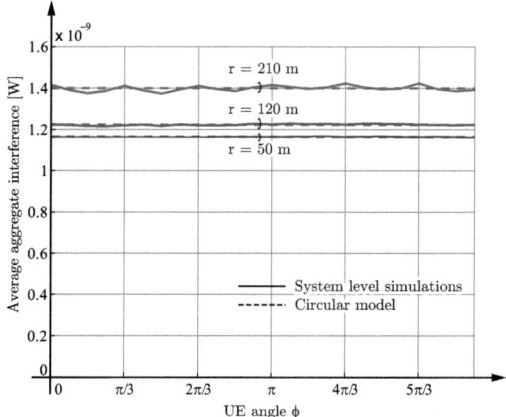

Figure 6.4.: Average aggregate interference power along the three UE circles in Figure 6.3. Solid curves refer to simulation results, dashed curves denote results from circular interference model.

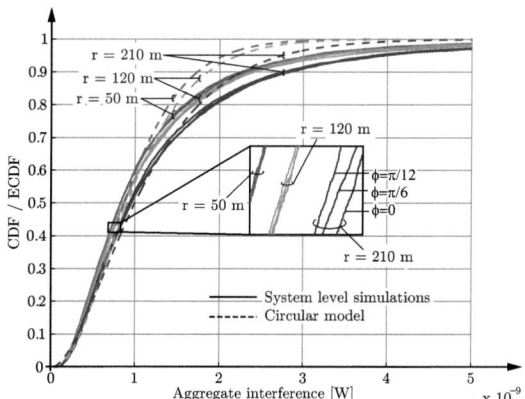

Figure 6.5.: Aggregate interference distributions at representative UE locations $r = \{50, 120, 210\}$ m and $\phi = \{0, \frac{\pi}{12}, \frac{\pi}{6}\}$, as marked by bold dots in Figure 6.3. Solid lines refer to ECDF curves from simulations, dashed lines denote Gamma approximations as obtained with circular model from Chapter 3.

Chapter 6. LTE-A System Level Simulations

Finally, the circular model from Chapter 3 is applied to approximate the aggregate interference distribution at a certain distance r by a Gamma RV. The first step consists in determining the parameters k_0 and θ_0 of the Gamma distribution $\Gamma[k_0, \theta_0]$ that models the composite fading (conf. Section 3.3). This is achieved by applying Algorithm 2. The intital values k_0' and θ_0' are obtained from Maximum Likelihood Estimation (MLE)[4]. Using a step size of $\Delta = 0.001$ and $N_{\text{iterations}} = 100$ yields a KS distance of 0.0512 between simulated- and approximated composite fading distribution. For comparison, employing only MLE achieves a KS distance of 0.0917.

Algorithm 2: Iterative algorithm for improving KS distance between empirical composite fading distribution and Gamma approximation. The term $F_\Gamma(x; k, \theta)$ denotes the CDF of a Gamma distribution with shape k and scale θ, respectively.

Data: empirical CDF of composite fading from simulations: $F_{\text{fading}}(x)$;
  initial shape- and scale parameter: k_0', θ_0';
  stepsize: Δ;
  number of iterations: $N_{\text{iterations}}$;
Result: shape- and scale parameter: k_0, θ_0;
set $k_0 = k_0'$ and $\theta_0 = \theta_0'$;
for $i = 1$ **to** $N_{\text{iterations}}$ **do**
  compute $\{k^*, t^*\} = \arg\min_{\{k,t\}} \sup_x |F_{\text{fading}}(x) - F_\Gamma(x; k, \theta)|$, with
  $k \in [k_0 - \Delta, k_0, k_0 + \Delta]$ and $\theta \in [\theta_0 - \Delta, \theta_0, \theta_0 + \Delta]$;
  if k^* *equals* k_0 **and** θ^* *equals* θ_0 **then**
   | break;
  else
   | set $k_0 = k^*$ and $\theta_0 = \theta^*$;
  end
end

Then, for each UE distance, the parameters of the aggregate interference distribution, $\hat{k}(r)$ and $\hat{\theta}(r)$, are calculated with (3.16) and (3.17), respectively. The corresponding CDF curves are depicted as dashed lines in Figure 6.5. It is observed that the approximated distributions slightly underestimate the occurrence of high interference values. In order to quantify the deviation from the simulated curves, the first row in Table 6.3 provides the KS distance for each UE location (r, ϕ). The values range from 0.05 at $r = 210$ m to 0.08 at $r = 50$ m.

For comparison, each simulated curve is also approximated by two further Gamma distributions. The first distribution adapts the circular model and estimates the composite fading by MLE, i.e., it employs the parameters k_0 and θ_0 that were used above to initialize Algorithm 2. The second

---

[4] MLE maximizes the *likelihood* $\mathcal{L}(k_0', \theta_0'|x) = f(x|k_0', \theta_0')$, where $f(\cdot)$ denotes a Gamma PDF with shape k_0' and scale θ_0', and x are the given outcomes.

Table 6.3.: KS distances between Gamma approximations and simulated ECDF curves at representative UE locations. For each r, the first two rows correspond to the Gamma approximation as obtained with the circular model. In the first row, composite fading is estimated with Algorithm 2, while in the second row it is assessed with MLE, respectively. The third row refers to the direct application of MLE on the distribution of the aggregate interference.

	$\phi = 0$	$\phi = \frac{\pi}{12}$	$\phi = \frac{\pi}{6}$
r = 50 m	0.0713	0.0768	0.0762
	0.1391	0.1491	0.1454
	0.0720	0.0797	0.0773
r = 120 m	0.0697	0.0659	0.0698
	0.1347	0.1393	0.1300
	0.0815	0.0708	0.0701
r = 210 m	0.0565	0.0496	0.0466
	0.1183	0.1289	0.1274
	0.0823	0.0828	0.0840

distribution is computed by applying MLE directly to the simulated aggregate interference. The corresponding KS distances are likewise listed in the second- and third row of Table 6.3 for each UE location (r, ϕ). The first observation is that Algorithm 2 considerably improves the performance of the circular model, such that it even exceeds pure MLE of the aggregate interference. Hence, the accuracy of the circular model crucially depends on the precision of the composite fading approximation. Secondly, the results of the MLE range from 0.07 to 0.08, indicating that the assumption of Gamma-distributed interference itself induces a systematic error.

In summary, the circular model achieves a remarkable accuracy of fit despite its simplicity, thus corroborating its applicability.

6.2.3. Validation of Asymmetric Interference Impact

In this section, SINR and spectral efficiency at eccentric UE locations are evaluated, validating results in Chapter 4. The system model is directly adopted from Section 6.2.1. The corresponding circular model from Section 4.1 encompasses one circle ($C = 1$) with radius $R_1 = 500$ m and six transmitters. According to Section 4.3, such model allows to *exactly* reproduce a regular grid model in terms of aggregate interference characteristics, provided that the composite fading follows a Gamma distribution. In order to omit the error, which is induced by the Gamma approximation as detailed in Section 6.2.2, *composite fading* is modeled by free space path loss and Rayleigh fading in both simulations and analysis (i.e., $G_{c,n} \sim \Gamma[1,1]$ in (4.1)).

Chapter 6. LTE-A System Level Simulations

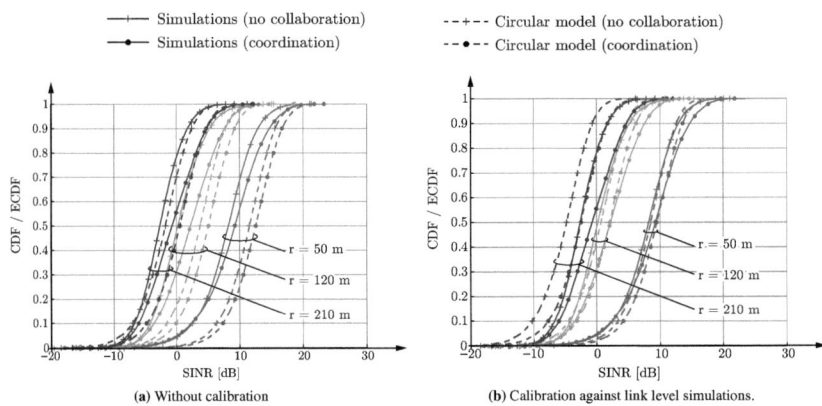

Figure 6.6.: SINR at various UE distances $r = \{50, 120, 210\}$ m and angle position $\phi = 0$, considering *no collaboration-* and *interference coordination* among the eNodeBs. The curves show results as obtained by system level simulations and the circular model from Section 4.4 with-(b) and without calibration (a), respectively.

The particular aim of this section is to verify results from Section 4.4. Accordingly, two scenarios, namely *no collaboration among transmitters* and *interference coordination* are defined. In the latter case, eNodeB 7 in Figure 6.3 does not contribute to the co-channel interference. From the representative UE locations, as specified in Section 6.2.1 and marked by bold dots in Figure 6.3, the particular interest of this section is on the angle position $\phi = 0$.

Figure 6.6(a) depicts SINR distributions for both collaboration schemes, comparing results from system level simulations and the circular model. It is observed that the SINR consistently deteriorates for a UE moving from cell-center to cell-edge. In accordance with Section 4.4, it is improved by interference coordination, with the largest gains being achieved at cell-edge. The simulated median values increase by 1.2 dB, 1.8 dB and 1.9 dB at the cell-center, middle of the cell and cell-edge, respectively. Furthermore, it is seen that the curves from the simulations are steeper, i.e., have a smaller variance than those obtained with the circular model. This is mainly caused by the fact that the simulator employs a ZF receiver and measures the *post-equalization SINR*. In order to more accurately capture the receiver characteristics, the first adaption of the circular model concerns the shape of the fading distribution. It is set to $k = 2$, i.e., $G_{c,n} \sim \Gamma[2,1]$.

In a practical system, performance is decreased by a variety of design constraints. Hence, the circular model has to be further calibrated against simulations [214]. This chapter employs

6.2. Homogeneous Macro Cellular Network

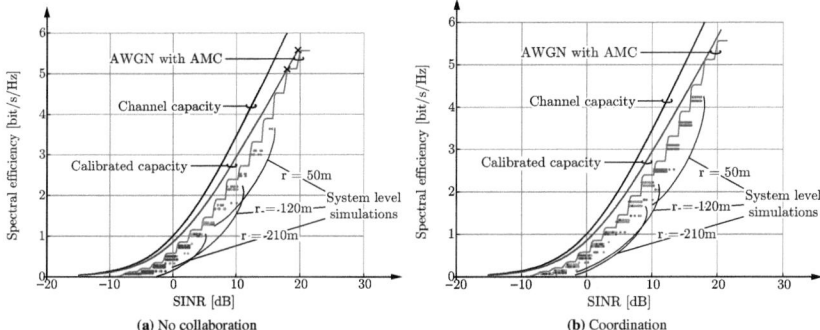

Figure 6.7.: Spectral efficiency [bit/s/Hz] versus SINR [dB] for baseline-(a) and coordination scheme (b). Dots refer to results from system level simulations at various user distance $r = \{50, 120, 210\}$ m. The curves refer to the Shannon channel capacity, the *calibrated* capacity (with $\alpha_\text{B} = 0.85$ and $\alpha_\text{SIR} = 0.5$) and the performance of a SISO LTE-A system over an AWGN channel employing AMC, as obtained from link level simulations.

the modified channel capacity formula $\tau(\gamma(r)) = \alpha_\text{B} \log_2\left(1 + \alpha_\text{SIR}\gamma(r)\right)$, as introduced in Section 4.4.3. The term $\gamma(r)$ denotes the SIR at distance r (note that in this section the UEs of interest have angle position $\phi = 0$), and α_B and α_SIR are freely adjustable calibration parameters, with $0 < \alpha_\text{B} \leq 1$ and $0 < \alpha_\text{SIR} \leq 1$.

Figure 6.6 depicts the simulation results from both baseline- and coordination scheme in terms of spectral efficiency versus SINR. The dots refer to the simulation results at various user distances $r = \{50, 120, 210\}$ m. For comparison, the figure also shows the *channel capacity* and the performance of a SISO LTE-A system over an AWGN channel. The latter is obtained from link level simulations and can expediently be used to predict the optimal performance of the system. Its discontinuous behavior stems from LTE-A's Adaptive Modulation and Coding (AMC) scheme [219]. It is observed that the results from the system level simulations lie below this curve, since they encompass the actual channel code performance [220]. In order to achieve reliable upper performance bounds, the link level results are employed as a reference for the calibration of the circular model, yielding $\alpha_\text{B} = 0.85$ and $\alpha_\text{SIR} = 0.5$, respectively[5]. Intuitively, the first term accounts for overhead, such as pilot symbols, while the second term represents the limits of the AMC in LTE-A [167, 221].

---

[5]The term α_B shifts the capacity curve in Figure 6.7, while α_SIR determines its scale. The calibration is carried out such that the curve is tangent to the link level results and achieves the maximum spectral efficiency in the scenario without collaboration, as denoted by '×' in Figure 6.7(a).

Chapter 6. LTE-A System Level Simulations

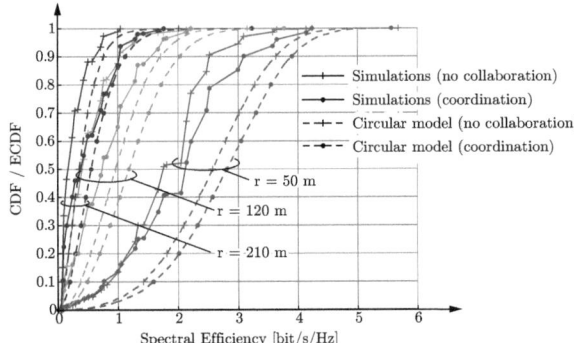

Figure 6.8.: Spectral efficiency [bit/s/Hz] at various UE distances $r = \{50, 120, 210\}$ and and angle position $\phi = 0$, considering *no collaboration*- and *interference coordination* among the eNodeBs. The curves show results as obtained by system level simulations and the circular model from Chapter 4 with $\alpha_B = 0.85$ and $\alpha_{SIR} = 0.5$, respectively.

The dashed curves in Figure 6.6(b) show the adapted SIR distributions. Except for $r = 210$ m, they exhibit a better fit than the uncalibrated curves in both shape and scale. The deviation at the cell-edge ($r = 210$ m) is partly compensated by overestimating the link level performance in the low-SINR regime (conf. Figure 6.7). The corresponding spectral efficiency distributions are obtained by applying (4.24) from Section 4.4.3[6]. They are depicted in Figure 6.7 together with the simulation results. It is observed that, despite the large number of simulated TTIs, the curves from the simulations exhibit a discontinuous behavior due to the AMC. As expected, the circular model provides reliable upper performance bounds that are tightest at $r = 210$ m.

In conclusion, the circular model from Section 4.1 is well suited to make a first-order prediction of the SINR- and spectral efficiency performance. The presented calibration against link level simulations has to be carried out only once for each MISO- or SIMO transmission scheme. This method forfeits a certain amount of accuracy while avoiding tedious ad-hoc calibration against each system level simulation run.

---

[6]This chapter employs the term *spectral efficiency* instead of *normalized rate*.

6.3. Two-tier Heterogeneous Cellular Network

In this section, two-tier heterogeneous cellular networks are investigated, encompassing eNodeBs on macro sites and femtocell BSs, respectively.

Although numerous system level simulation campaigns have been carried out, the utilized system models such as the dual-stripe- and the 5 × 5 approach from [202] or other customized setups such as [15, 222–226] are mostly too specific to *systematically* investigate the impact of a *femtocell enhancement* on the existing macro cellular deployment. On the other hand, analytical work is commonly evaluated in terms of capacity and cannot directly be transferred to achievable throughput due to highly idealistic setups.

This section introduces a system model, which enables to analyze the impact of UE distribution, femtocell deployment density and femtocell isolation on the performance of a two-tier heterogeneous cellular network in a *systematic* manner. Moreover, it allows to discuss results from Chapter 5. In contrast to all previous considerations, the focus of this section is on *network-wide* performance metrics, i.e., a *global*- rather than a *user-centric* point of view.

6.3.1. System Model

The macro cellular setup comprises a central site and one tier of hexagonally arranged neighbors, as illustrated in Figure 6.9. Each site employs three eNodeBs, which are equipped with one sector antenna each. The antennas are arranged at a spacing of $2\pi/3$. Their radiation pattern is referred from [167] and is specified as

$$A(\theta) = -\min\left[12\left(\frac{\theta}{\theta_{3\text{dB}}}\right)^2, A_m\right], \quad -\pi \leq \theta \leq \pi, \quad (6.1)$$

where $\theta_{3\text{dB}} = \frac{65}{180}\pi$ and $A_m = 20$ dB. Applying maximum-SINR-based UE association, such setup yields hexagonally shaped eNodeB coverage-regions, as illustrated in Figure 6.9. Hereinafter, these regions are referred to as *macro sectors*.

After establishing the macro deployment, N_C circularly-shaped buildings of radius $R_I = 25$ m are uniformly distributed within each macro sector such that their footprints *do not overlap* each other. Along the lines of Section 5.2, a point is denoted as *indoors*, if it is covered by a building, and *outdoors* otherwise.

Each building hosts N_U UEs, which are uniformly distributed within an annular region of inner radius 5 m and outer radius 25 m around the center of the building. This procedure is equivalent

Chapter 6. LTE-A System Level Simulations

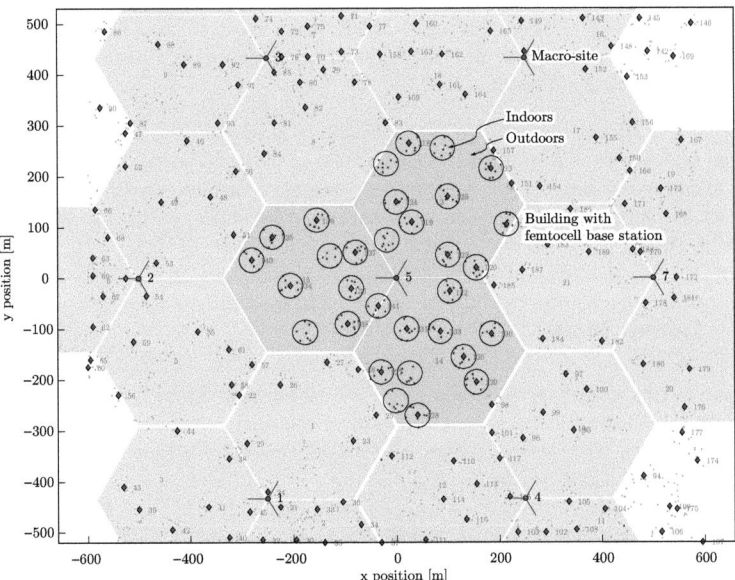

Figure 6.9.: Urban two-tier heterogeneous cellular network. UEs are located within annular regions around the centers of the randomly distributed buildings. Buildings are served by a femtocell BS with occupation ratio $\eta = 0.8$. Indoor- and outdoor environment are separated by wall penetration loss L_W.

Table 6.4.: Specific simulation parameters for two-tier heterogeneous cellular network.

Parameter	Value
eNodeB deployment	three eNodeBs per macro site, $2\pi/3$ spacing
eNodeB antenna gain	$A(\theta) = -\min\left[12\left(\frac{\theta}{\theta_{\text{3dB}}}\right)^2, A_m\right], -\pi \leq \theta \leq \pi$
Femtocell BS transmit power	$P_S = 26\,\text{dBm}$
Femtocell backhaul	unlimited, no delay
Femtocell access mode	{OSG, CSG}
Femtocell antenna gain	0 dB omni-directional

to generating *UE hot-spots* according to an independent cluster process[7]. In this case, the parent process is constituted by the distribution of the building centers. A *UE-cluster* is formed by the UEs of a given building. In total, there are $N_S = N_C N_U$ UEs in each macro sector, hereafter denoted as *sector UEs*. By keeping N_S constant, the parameters N_C and N_U adjust the *degree of clustering*, also referred to as *extent of clustering* or *level of inhomogeneity* [228].

Femtocell BSs are deployed at the centers of buildings and equipped with omni-directional antennas. Their occupation ratio η is tuned by N_F/N_C, with $0 \leq N_F \leq N_C$ (conf. Section 5.2.2). The parameter N_F denotes the number of randomly chosen buildings per macro sector, which are equipped with a femtocell BS. Figure 6.9 shows a snapshot with $N_F = 8$ and $N_C = 10$, respectively. Distributing the UEs within an annular region around the building centers guarantees a certain minimum distance to the femtocell BSs.

Both OSG- and CSG mode are considered. In the OSG case, a femtocell BS serves all UEs within its coverage area[8], whereas in the CSG mode, only the UEs of the corresponding cluster are allowed to attach. UEs associated with an eNodeB from a macro site are denoted as *macro UEs*, UEs attached to a femtocell are referred to as *femto UEs*, respectively. Macro- and femtocell tiers are assumed to be uncoordinated and employ universal frequency reuse, i.e., reuse-1, thus representing a worst-case scenario in terms of interference.

Signal propagation in- and out of a building is modeled by a constant wall penetration loss L_W. Depending on whether a signal originates from an eNodeB on a macro site or a femtocell BS, one of the following two path loss models is applied:

- *eNodeB on macro site*: The path loss model is referred from [167, subclause 4.5.2][9] and depicted as dashed line in Figure 6.10.

---

[7]See, e.g., [227] for nomenclature and further details on cluster processes.
[8]UE association regions are calculated according to a maximum SINR criterion.
[9]Exemplifying from [167], for a carrier frequency of 2.14 GHz and a BS antenna height of 15 m above average rooftop level, $\ell(R)_{\text{[dB]}} = \min(-A_{\text{[dB]}} - 128.769\,\text{dB} - 37.6\log_{10}(R), -70)$, where $A_{\text{[dB]}}$ is the antenna gain and R is the distance in kilometers.

Chapter 6. LTE-A System Level Simulations

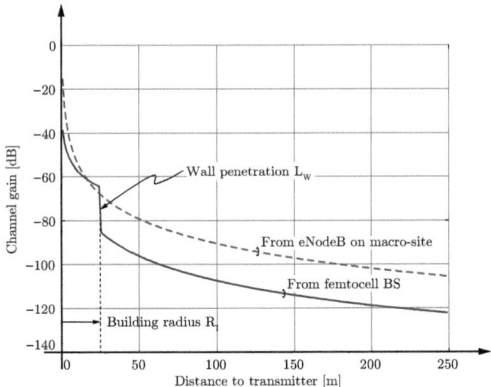

Figure 6.10.: Distance dependent channel gain from eNodeB on macro site and femtocell BS.

- *Femtocell BS*: A dual-slope model is applied (solid line in Figure 6.10). Within the building, the propagation model as specified in [229, subclause 5.2][10] is employed. At distance R_I the signal is attenuated by the wall penetration loss L_W. For distances larger than R_I, again the propagation loss model from [167, subclause 4.5.2] is utilized.

Small scale fading and shadowing are modeled by time correlated Rayleigh- and spatially-correlated log-normal RVs, respectively. The simulation parameters are summarized in Tables (6.1) and (6.4), respectively.

The introduced system model serves as a basis for the subsequent investigations on altering UE distribution, building characteristics and urban environment. The presented simulation results are obtained by taking into account the three sectors of the central macro site and averaging over 100 scenario snapshots and 100 TTIs per snapshot, respectively. *Network-wide* performance is represented by *sector-wise* metrics.

6.3.2. Urban Two-tier Heterogeneous Cellular Network

In this section, the performance of a typical indoor-UE in an urban two-tier heterogeneous cellular LTE-A network is evaluated. The target is to verify results from Chapter 5. Simulations

---

[10] Exemplifying from [229], $\ell(R)_{[\text{dB}]} = \min(-98.46\,\text{dB} - 20\log_{10}(R), -45)$, where R is the distance in kilometers.

are carried out with the system model from Section 6.3.1, which, however, differs from the model in Section 5.2 in two major aspects:

- Macro sites are arranged according to a hexagonal grid, as depicted in Figure 6.9. This setup guarantees a certain macro sector size that is necessary to carry out systematic simulations, as detailed in Section 6.3.1.

- The characteristics of the urban topology, such as building density, are incorporated into the standard deviation of the log-normal RVs which represent the shadowing. Typical values in literature range from 6 – 10 dB [195, 230–236]. The rationale for this model is to emphasize the difference to the exponential law in (5.5).

In accordance with Chapter 5, performance is evaluated for two wall penetration losses $L_W = \{-10, -30\}$ dB, and two femtocell occupation ratios $\eta = \{0.2, 0.8\}$, respectively. Figure 6.11 shows the average spectral efficiency of a *typical indoor UE* plotted over the shadow fading standard deviation. It is observed that

- The standard deviation of the shadow fading has almost no effect on the performance. Thus, the impact of the urban environment topology is completely overlooked with the log-normal model. Moreover, the model neglects the differentiation between LOS- and NLOS BSs, as investigated in Section 5.5. The importance of the latter is manifested through its inclusion into Rel. 12 of the LTE-A standard [185].

- The performance in OSG- and CSG-mode considerable deviate at high femtocell occupation ratio ($\eta = 0.8$) and low wall penetration ($L_W = -10$ dB). This is caused by the fact that, in the OSG case, UEs in a building without femtocell-BS can associate with a femtocell in a close-by building. In contrast, CSG operation forces these UEs to associate with the eNodeB on the macro site. Moreover, they will receive severe interference from the nearby femtocell-BS. This aspect is further investigated in Section 6.3.4 as it is not considered in the model of Chapter 5.

For comparison, a numerical evaluation of (5.17) with $P_c(\delta|r)$ from (5.16) is carried out[11], assuming that the users are uniformly distributed within an annulus of inner radius 5 m and outer radius $R_I = 25$ m, respectively. The curves are computed with the settings $P_S/P_M = 10^{-2}$ and $\mu_M = 4.61 \cdot 10^{-6}$ m^{-2} (according to the BS density in a hexagonal grid with an inter-site distance of 500 m), and the calibration parameters $\alpha_B = 0.85$ and $\alpha_{SIR} = 0.5$ from Section 6.2.3. According to Section 6.3.1, outdoor- and indoor path loss are specified as $\ell_O(R) = \min(10^{-7}, 1/10^{1.597} R^{-3.76})$ and $\ell_I(r) = \min(10^{-4.5}, 1/10^{3.846} r^{-2})$, respectively. Figure 6.12 shows the results.

---

[11] Since the simulator does not distinguish LOS- and NLOS BSs, it is refrained from employing the more elaborated model from Section 5.5.

Chapter 6. LTE-A System Level Simulations

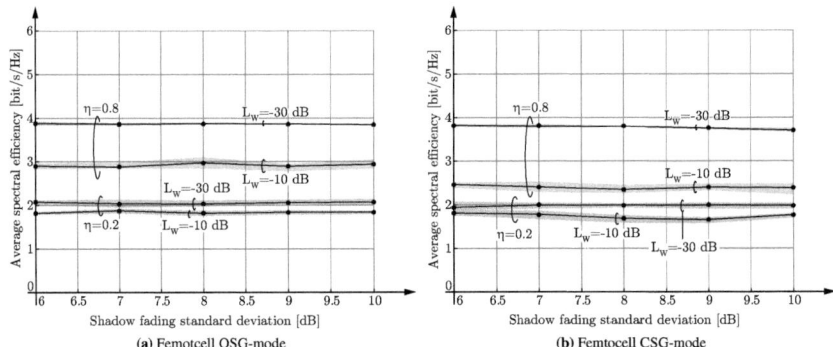

Figure 6.11.: Average spectral efficiency [bit/s/Hz] of typical indoor UE. Results are provided for femtocells operating in OSG- and CSG mode, respectively. Performance is evaluated for two wall penetration losses $L_W = \{-10, -30\}$ dB, and two femtocell occupation ratios $\eta = \{0.2, 0.8\}$, respectively. The shaded regions denote 95 % confidence intervals.

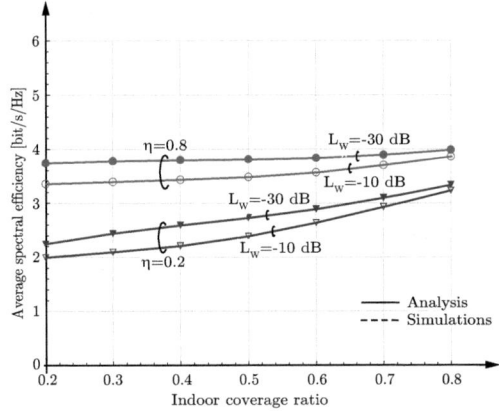

Figure 6.12.: Spectral efficiency [bit/s/Hz] of typical indoor user versus area ratio that is covered by buildings. Curves are obtained by numerical evaluating the theoretical model from Chapter 5 and applying the calibration parameters $\alpha_B = 0.85$ and $\alpha_{SIR} = 0.5$. Curves are depicted for varying small cell occupation probability η and wall penetration loss L_W.

It is observed that, in agreement with Section 6.2.3, the theoretical model tends to overestimate the performance due to the conservative calibration against the ideal system performance. The deviation is particularly pronounced at high indoor area coverage, where the building blockage provides a safeguard against interference. It is further remarkable that, in contrast to the simulation results, the analytical curves show a similar performance improvement when altering the wall penetration from $L_W = -10\,\text{dB}$ to $L_W = -30\,\text{dB}$ for *both* $\eta = 0.2$ and $\eta = 0.8$. This is caused by the fact that the model only takes into account interference from *neighboring* femtocells.

In conclusion, the model from Chapter 5 enables more subtle insights on the effects of building blockages than log-normally distributed shadowing, including the safeguard against interference as well as the decreasing impact of LOS BSs, which were not included in the simulations due to the novelty in the 3GPP standard [185]. On the other hand, it misses aspects of OSG and CSG-operation. The subsequent sections complete the picture by systematically evaluating *network-wide* performance. In the remainder of this chapter, shadowing is modeled by spatially correlated log-normal RVs with $8\,\text{dB}$ standard deviation.

6.3.3. User Hot Spot Scenarios

In this section, the impact of *UE clustering* on the *global* performance of femtocell-enhanced macro cellular networks is investigated. Current research mainly focuses on the positioning of the transmitters while users are commonly considered uniformly distributed [7, 19, 21, 26, 49, 119, 237, 238]. However, femtocell BSs are most effectively deployed at user hot-spots [6, 239]. Hence, appropriate models for the user distribution are essential to investigate the *performance limits* of a heterogeneous network. Based on [79], the contributions of this section are:

- A system model is presented, which enables to explicitly identify the effects of altering the *degree of UE clustering*.
- The importance of a fairness metric is stressed, as it is often disregarded in literature. By means of sum throughput, it is shown that a network-wide performance metric provides only limited view on the UE performance, since it conceals the distribution of the individual values.

In this section, the system model from Section 6.3.1 is exploited without separating indoor- and outdoor environment, i.e., $L_W = 0\,\text{dB}$. For a fair comparison of different UE distributions, the total amount of UEs per sector, N_S, is kept constant over all simulations. The degree of clustering is tuned by the parameters N_C and N_U, i.e., the number of UE clusters per sector and the number of UEs per cluster, such that $N_S = N_C N_U$.

Chapter 6. LTE-A System Level Simulations

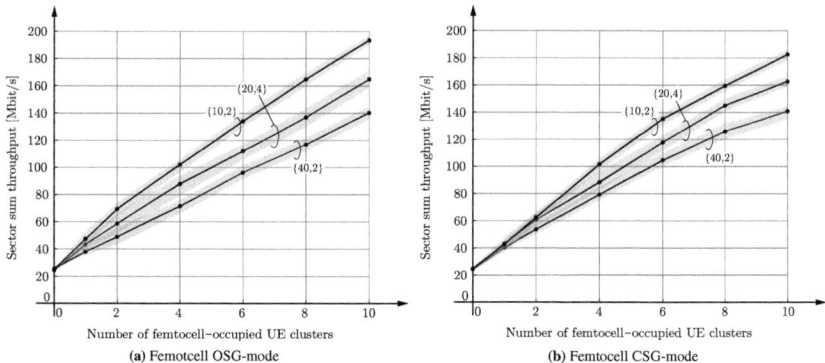

Figure 6.13.: Sector sum throughput [Mbit/s] over number of femtocell-occupied UE clusters.

The parameter settings for simulations are summarized in Tables (6.1) and (6.4), respectively. *Sector-wise sum throughput* is employed as a *global* performance metric. It is calculated by accumulating the throughput values of all N_S sector UEs, i.e., including macro- and femto UEs. Such metric is of particular interest for network providers when planning a femtocell roll-out.

The results are depicted in Figure 6.13. Three scenarios are investigated, ranging from a low- to a high degree of UE clustering. The corresponding $\{N_C, N_U\}$-tuples are specified as $\{40, 2\}$, $\{20, 4\}$ and $\{10, 8\}$, respectively. It is observed that in all three cases the sum throughput strictly increases with additional femtocells, thus confirming the claim in [7]. The results however reveal that the efficiency of the femtocell operation considerably depends on the degree of clustering. The sum throughput increases steepest in a UE hot-spot scenario, and lowest in a close to uniform UE distribution. Remarkably, the curves show a slight saturation effect when increasing the number of femtocell BSs. This indicates that the femtocell deployment density does not perfectly compensate for the additional interference.

Sum throughput is suitable to measure the global performance of a femtocell-enhanced network. However, it conceals possible performance imbalances between the individual UEs. Therefore, measures for the distribution of the throughput values are imperative. In this section, Jain's fairness index is employed. For a given macro sector, it is expressed as

$$\mathrm{JFI}(\mathbf{t}) = \frac{\left(\sum_{i=1}^{N_S} t_i\right)^2}{N_S \sum_{i=1}^{N_S} t_i^2}, \quad (6.2)$$

6.3. Two-tier Heterogeneous Cellular Network

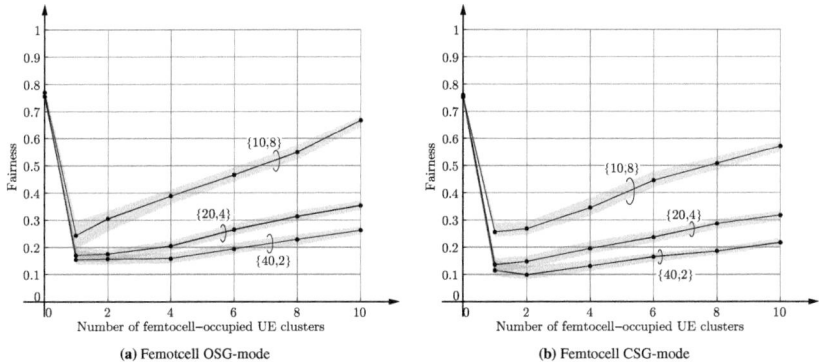

Figure 6.14.: Jain's fairness index over number of femtocell-occupied UE clusters.

where t_i denotes the throughput as achieved by UE i.

Figure 6.14 depicts the fairness index plotted over the number of femtocell occupied UE clusters per macro sector. In a sparse femtocell deployment (i.e., low number of femtocell BSs), only few UEs achieve high throughput due to their vicinity to the femtocell BSs. The remaining UEs are attached to the eNodeB on the macro site, experiencing additional interference from the femtocells. Thus, low fairness is observed. The index increases with the number of employed femtocell BSs. At full femtocell occupation ($\eta = 1$), it lies below the reference case of a plain macro cellular network without femtocells ($\eta = 0$) for both OSG- and CSG mode. In accordance with the sum-throughput results, highest fairness is achieved at the largest extent of UE clustering.

Hence, femtocells are most efficiently deployed in scenarios with a high degree of UE clustering. Motivated by the low fairness values, macro- and femto UE performance are evaluated separately in the next section. The particular focus is on the effect of isolating UEs in indoor areas from the outdoor environment by wall penetration loss.

6.3.4. Sensitivity on Femtocell Deployment Density and -Isolation

In this section, the impact of femtocell deployment density and -isolation on the downlink-performance of an LTE-A network is investigated. *Isolation* is defined as the separation between indoor- and outdoor environment by wall penetration loss.

101

Chapter 6. LTE-A System Level Simulations

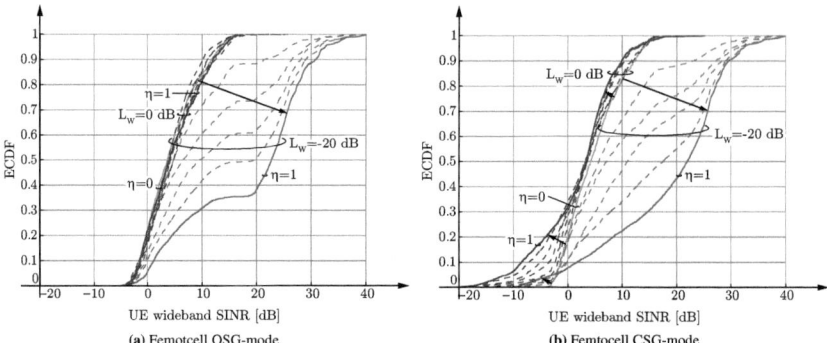

Figure 6.15.: SINR ECDFs for various femtocell-occupation ratios $\eta = \{0, 0.2, 0.4, 0.6, 0.8, 1\}$ and wall penetration losses $L_W = \{0, -20\}$ dB, respectively. Arrows denote the direction of increasing η. The case $\eta = 0$ serves as a baseline, corresponding to a macro cellular network without femtocells.

Referring to my work in [80], the contributions of this section are:

- A system model is introduced, which allows to explicitly analyze the effects of varying femtocell density and -isolation.

- The individual UE performance, which would be concealed by network-wide performance metrics (conf. Section 6.3.3), is assessed by separately investigating macro- and femto UEs.

The system model which is largely based on the setup in Section 6.3.1. *High-* and *no-isolation* scenarios are investigated, corresponding to $L_W = -20$ dB and $L_W = 0$ dB, i.e., the worst case in terms of interference, respectively. According to results in Section 6.3.3, efficient balancing of UE throughput is only possible in scenarios with a high degree of UE clustering. Therefore, $N_C = 10$ and $N_U = 8$ in the remainder of this section.

The first metric of interest is the wideband SINR. It is defined as the ratio of the average receive power from the serving cell and the average aggregate interference from other cells plus noise [240]. Figure 6.15 depicts the corresponding ECDF curves for OSG- and CSG mode, considering both high- and no isolation of the indoor areas. The curves are computed from the average wideband-SINR values of the N_S sector UEs. Various femtocell-occupation ratios $\eta = \{0, 0.2, 0.4, 0.8, 1\}$ are evaluated. Arrows indicate the directions of increasing η from 0 to 1. The case $\eta = 0$ serves as a baseline, representing a macro cellular network without femtocells.

6.3. Two-tier Heterogeneous Cellular Network

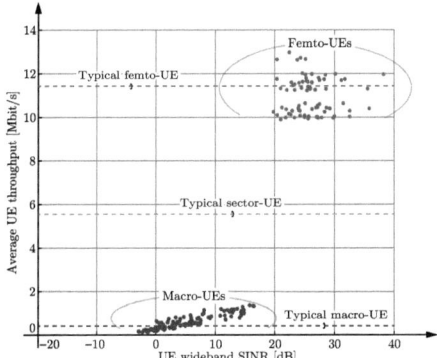

Figure 6.16.: Snapshot of individual average-UE-throughput values [Mbit/s] over wideband SINR [dB] for wall penetration loss $L_W = -20\,\text{dB}$, femtocell occupation ratio $\eta = 0.6$ and OSG mode. The dashed lines denote throughput values of the *typical* macro-, femto- and sector UE as referred from Figure 6.17(a), respectively.

The major observations are

- High- and no-isolation scenarios exhibit significantly different characteristics. On the one hand, for $L_W = -20\,\text{dB}$ the SINR almost consistently improves with increasing η. On the other hand, for $L_W = 0\,\text{dB}$ it hardly deviates from the baseline in the OSG case, while deteriorating in the CSG scenarios. These results indicate the system's high sensitivity to fluctuations of the femtocell isolation.

- The step-like behavior of the curves, which is particularly pronounced in OSG scenarios at high isolation, indicates a severe imbalance between UEs with good- and UEs with bad channel conditions. Figure 6.16 depicts a snapshot of individual average-UE-throughput values over SINR for $L_W = -20\,\text{dB}$, $\eta = 0.6$ and OSG mode. It allows to unambiguously identify the two groups as macro- and femto UEs, thus motivating their *separate* investigation.

Figure 6.17 shows throughput values as achieved by the *typical* macro-, femto- and sector UE, respectively. The results are obtained by averaging over the individual throughput values of the corresponding UE class. The axis of abscissas shows the femtocell occupation ratio η. It is found that

- The throughput of the typical sector UE monotonically increases with larger η and achieves its maximum at full femtocell occupation, i.e., $\eta = 1$.

103

Chapter 6. LTE-A System Level Simulations

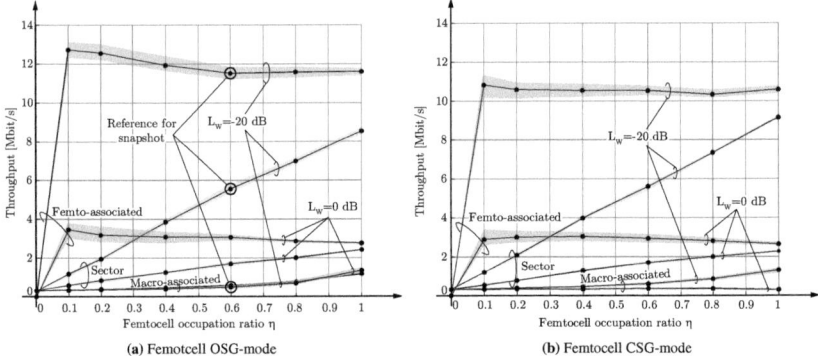

Figure 6.17.: Throughput of typical macro-, femto- and sector UE [Mbit/s] over femtocell occupation ratio η for $L_W = \{0, -20\}$ dB. Shaded regions denote 95 % confidence intervals. The circled points in (a) corresponds to the straight dashed lines in Figure 6.16.

- The performance of a typical femto UE is considerably higher for high isolation ($L_W = -20$ dB) than for no isolation ($L_W = 0$ dB). The latter might serve as a *warning* scenario for network providers, considering open doors and windows throughout the whole scenario.

- In agreement with the results in Section 6.3.3 and the observations in the current section, the throughput of the typical sector UE considerably deviates from the typical macro- and femto-UE performance. This becomes particularly clear in OSG scenarios with $L_W = 0$ dB, where the throughput of the typical sector UE is enhanced while the performance of the typical macro UE deteriorates.

- Except for the aforementioned case, the throughput of the typical macro UE generally improves for increasing η. This is caused by the fact that handing off a growing amount of macro UEs to the femtocells compensates for the harm of additional interference.

Mapping the circled throughput values in Figure 6.17(a) to Figure 6.16 (straight dashed lines) substantiates the observation that the *individual* UE either performs much better or much worse than the *typical* sector UE. On the other hand, typical macro- and femto UE throughputs provide reasonably accurate indicators for the actual performances.

Hence, these two metrics bridge the gap between individual- and sector-wise performance. Figure 6.18 depicts typical femto- versus typical macro UE throughput. Each point represents a

6.3. Two-tier Heterogeneous Cellular Network

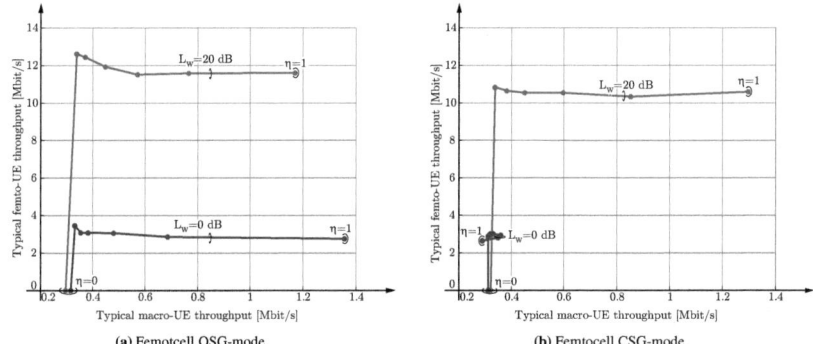

Figure 6.18.: Typical femto- versus typical macro UE throughput [Mbit/s]. Each point corresponds to a certain femtocell occupation ratio $\eta \in \{0, 0.1, 0.2, 0.4, 0.6, 0.8, 1\}$. Lines connect throughput tuples of successive η-values.

certain femtocell occupation ratio η, with lines connecting successive values. This depiction is conceptually similar to a *capacity-* or *throughput trade-off region* with the competing axes being typical macro- and typical femto UE throughput [43, 241], respectively. The following observations are equivalently obtained from Figure 6.17, but become much clearer from Figure 6.18:

- The typical femto UE throughput is almost constant for $\eta > 0$. High- and no-isolation scenario differ by a factor of about 3.8 in OSG- and about 3.5 in CSG-mode, respectively.

- In OSG mode the average macro UE throughput grows with increasing η at $L_W = 0$ dB, even exceeding the performance at $L_W = -20$ dB. In the CSG case, the throughput tuples exhibit a loop, yielding the lowest throughput of the typical macro UE at full femtocell occupation. Thus, operation in CSG mode is more resilient to fluctuations of L_W.

In conclusion, the results in this section confirm the assumption from Section 6.3.3 that the imbalance among individual UE performance values mainly results from the difference between macro- and femto UEs. On the other hand, performances within the corresponding UE class turned out to be relatively similar and motivated to introduce the notions of *typical macro-* and *typical femto UE*. Depicting their throughput values in a similar manner to a capacity region revealed that *femtocell isolation* mainly scales the throughput of the typical femto UEs while *femtocell density* and *femtocell access mode* predominantly affect the performance of the typical macro UE.

6.4. Summary

This chapter complements observations from the previous chapters by LTE-A system level simulations. The first part validates the applicability of the theoretical models as presented in Chapters 3 to 5 in a realistic environment. It is shown that the circular model from Chapter 3 enables an accurate prediction of the interference statistics in a hexagonal grid scenario. Deviations from the simulation results mainly stem from the inaccurate approximation of the composite fading. The remainder of the approximation error is caused by the assumption of Gamma distributed aggregate interference itself. Next, it is demonstrated that the circular model from Chapter 4 enables a convenient first-order prediction of the actual performance, which is particularly accurate in terms of SINR but slightly overestimates the achievable spectral efficiency. The latter is caused by a conservative calibration against link level simulation results, and is also partly responsible for the discrepancies between the urban two-tier model from Section 6.3.2 and the system level simulation results. The major deviations in this scenario, however, result from the use of the log-normal shadowing model, which is shown to conceal essential effects such as interference protection and the presence of LOS links, which gained momentum not until Release 12 of the LTE-A standard.

The second part of the chapter places particular emphasize on global performance metrics. It is shown that, in terms of sum-throughput and fairness, femtocells are most efficiently deployed in scenarios with a high degree of UE clustering. However, the performance values of the individual UEs exhibit a large variance, which mainly originates from the difference between macro- and femto UEs. Separately calculating their *typical* performance and plotting them in the manner of a capacity region yields clear insights on the mutual behavior at increasing femtocell density. It is observed that femtocell isolation mainly impacts the throughput of the typical femto UE while the performance of the typical macro-UE is essentially determined by the femtocell density and the femtocell access mode.

Self critically analyzing the results of this chapter, the following issues may be worth rethinking. In Section 6.2.2 it is observed that MLE is not optimal in terms of KS statistics. The accuracy of the composite fading approximation is improved by Algorithm 2. However, it is not shown whether the proposed method achieves the best possible result. Hence, it would be interesting to undertake further investigations on approaches that are optimal in terms of KS statistics. Moreover, it is observed that the assumption of Gamma-distributed aggregate interference itself induces a systematic approximation error. However, as already stated in Chapter 3, there is no known method to determine the most suitable statistics for this task.

In Section 6.2.3, it is demonstrated that the calibration of the spectral efficiency against results from link level simulations achieves relatively loose upper performance bounds. It would be

6.4. Summary

interesting to scrutinize the accuracy improvement by ad-hoc adaption against the actual system level simulation results.

In Section 6.3.2, it is observed that the log-normal shadowing model might neglect essential effects of an urban environment topology. However, these results are difficult to reproduce in practice, as it is hardly feasible to measure the performance of a *typical* indoor user in a multitude of cities with the same characteristics. Moreover, such scenarios are far too complex for link-level simulations. The verification of the model might become possible with the adoption of 3GPP's 3D channel model [185] in the Vienna LTE-A simulator, which features LOS- and NLOS links as well as location-dependent shadowing.

Similarly, parameterizing the UE-cluster model from Section 6.3.3 with real-world data will be difficult to achieve in practice. This is caused by the fact that measuring the degree of inhomogeneity is an ongoing topic in research, mainly due to the lack of a definite mapping [228, 242].

Results in Sections 6.3.3 and 6.3.4 indicate that the *optimal deployment density* of femtocells would be the occupation of *all* UE clusters. However, except for additional interference, the employed metrics do not account for any other costs of a femtocell deployment, such as energy expenditure and frequency of hand-overs. Hence, it would be instructive to reconsider the results under the terms of a multi-objective optimization problem.

Section 6.3.4 introduces the notions of *typical macro-* and *typical femto UE*, respectively. While this concept is applicable in two-tier networks, it has to be studied whether it holds true for an arbitrary number of tiers.

In this chapter, UE association is based on a maximum-SINR metric. In a heterogeneous network with potentially massive differences between nominal cell sizes, *load* becomes another important criterion [7]. Thus, it would be insightful to reconsider the simulations with load-balancing measures such as cell-range extension.

Chapter 7.

Conclusions

Within the last decade, the intense demand for mobile data traffic has exceeded even the most aggressive predictions. Due to the ubiquitous use of the Internet and the rapid adoption of novel devices such as smart phones and tablet computers, this trend is likely to persist. Moreover, consumers have become accustomed to ever-present wireless access and low-price rates, pushing network providers, standardization bodies as well as researchers to devise sustainable- and economically viable solutions for entering the so called *era of capacity*.

One of the most promising approaches is the reduction of competing users, hence decreasing the size of the cells. Nevertheless, practical feasibility requires such small cells to be deployed on demand rather than on a global scale. Therefore, cellular networks are becoming increasingly heterogeneous, forcing the wireless community to substantially rethink classical concepts of network modeling and -design.

The main contribution of this thesis is the development of system models for interference analysis and systematic simulation of LTE-A compliant heterogeneous cellular networks. While the proposed concepts claim analytical tractability and generality of the obtained results, their practical relevance is considered equally important. The following section summarizes the main contributions of this thesis.

7.1. Summary of Contributions

The first part of this thesis is devoted to a *user-centric* investigation of aggregate interference. Chapter 3 presents a *circular model*, which simplifies interference analysis of regular-grid deployments. It accounts for the fact that the hexagonal grid, despite its popularity and practical relevance, has rather undesirable properties for a rigorous mathematical evaluation. On the other hand, well-planned macro-sites are not expected to disappear from heterogeneous network

Chapter 7. Conclusions

topologies in the medium-term. The proposed approach accurately models aggregate interference statistics at arbitrary user locations by a single Gamma RV. It thus enables to characterize the entire macro-tier by only few key parameters. Simulations in a realistic environment reveal that deviations from the actual interference distribution are mainly caused by the imprecise approximation of the composite fading.

In Chapter 4 in this thesis, an *enhanced circular model* is proposed, which enables to represent arbitrary network topologies by a *well-defined* symmetric circular structure such that the original interference statistics are preserved. This is achieved by employing angle-dependent power profiles along the circles. The chapter presents a heuristic scheme for mapping arbitrary heterogeneous BS deployments. Although not claimed to be optimal, it achieves to accurately capture thousands of interferers by several tens of nodes. Considering the model's finite number of transmitters, the chapter also presents a new finite sum representation for the PDF of the sum of Gamma RVs with integer-valued shape parameter. The resulting framework enables to identify the nodes which principally shape the interference distribution at arbitrary user locations within the central cell, yielding the most favorable candidates for BS coordination and -cooperation. What is more, it allows to demonstrate the potential gains of such collaboration schemes. LTE-A simulations corroborate the model's applicability for a first-order performance prediction. Deviations mainly originate from the conservative calibration against link level simulations.

In Chapter 5, the modeling paradigm is shifted from symmetric structures to random spatial distributions. In comparison to existing stochastic models, the main contribution is the embedding of the network into an *urban environment topology*, which determines the characteristics of both signal propagation and BS deployment. The problem of interference asymmetry is resolved by a *virtual building approximation*. Particular emphasis is placed upon the typical indoor user, revealing effects such as a safeguard against interference or the vanishing impact of LOS BSs and wall penetration at high building densities, which have been overlooked by existing models. This is confirmed by LTE-A system level simulations, which exhibit unvaried performance when altering the standard deviation of the log-normally distributed shadowing. Thus, the proposed model enables to detect essential effects of an urban environment.

The final part of the thesis complements the *user-centric* considerations of the first part with *network-wide* investigations. Chapter 6 presents results from LTE-A system level simulations. The first main contribution is a system model that enables to systematically scrutinize the impact of *UE clustering, femtocell deployment density* and *-isolation* on an existing macro-cellular network. Secondly, awareness is raised on the limited reliability of global performance metrics. It is observed that a femtocell deployment achieves the best performance in terms of sum throughput and fairness, when the UEs are highly clustered and all clusters are occupied by a femtocell BS. At high femtocell density, the results exhibit a saturation effect rather than a linear increase as predicted in theory [118]. The strong dependence on the degree of clustering provides an impetus to rethink metrics such as area spectral efficiency $[\mathrm{Mbit/s/Hz/m^2}]$, which gained momentum with the emergence of heterogeneous networks but assumes that users and load are evenly spread in space. A second major finding is the large variance among individual UE performance values, mainly arising from the gap between macro- and femto-UEs. The notions of *typical macro-* and *typical femto-UE* are proposed. Depicting the corresponding throughput values in the manner of a capacity region reveals that the typical femto UE performance is mainly dependent on the femtocell isolation while the typical macro UE performance is primarily determined by the femtocell density and -access mode.

7.2. Open Issues and Outlook

Notwithstanding the careful design and evaluation of the proposed models in this dissertation, there are still some issues left for further investigations so as to improve their practical relevance.

The two presented circular interference models largely rely on the assumption of Gamma-distributed composite fading. LTE-A system level simulations have revealed that this approximation may cause the greatest uncertainty in the course of calculation. Hence, it has to be scrutinized whether there are more suitable probability distributions for this task. A primal requirement is the existence of a sum distribution which typically goes hand in hand with a low amount of adjustable parameters. The second major source of divergence between theoretical models and simulations are the practical design constraints of LTE-A. Hence, their impact has to be carefully identified in further investigations, in order to improve the accuracy of the calibration.

The great weakness of the presented model for urban environment topologies is the negligence of reflections and diffractions. Further effort has to be made to capture these effects in a mathematically tractable manner. Apart from the practical design constraints, it is not clear for the moment, whether the observed discrepancies between the model and LTE-A simulations stem from this simplification, or the inaccuracy of the log-normal shadowing. A further restriction of this work arises from solely evaluating the performance of a typical *indoor* user. Findings from LTE-A simulations indicate that it is rather the performance of a typical *tier* user, which is crucial for the understanding of the network characteristics, thus yielding in interesting task for further analysis. Another issue that is not sufficiently researched yet, is whether the notions of *typical macro-* and *typical femto-user* hold true for an arbitrary number of tiers. Hence, further investigations have to be directed towards enhanced system models, which enable to

Chapter 7. Conclusions

systematically evaluate networks with multiple tiers and allow to draw inferences about the typical tier user.

This thesis is based on the fundamental assumption of two-dimensional scenarios. With the recent release of a three-dimensional channel model for the study of elevation beamforming and Full-Dimension MIMO [185], 3GPP has made a clear statement for the future of wireless network modeling. A considerable effort should be directed towards augmenting the existing models by a third dimension. Moreover, the models in this thesis only enable to account for SISO and SIMO/MISO beamforming by altering the shape parameter of the Gamma distribution. Since future wireless cellular system will heavily rely on MIMO transmissions [185], their support also yields an important topic for further work.

7.3. Conclusion

The presented models in this dissertation resolve challenging issues from the past and provide new concepts for the design and analysis of future heterogeneous cellular networks on system level. The associated simulation study promotes a systematic- and reproducible methodology, which is an absolute necessity in scientific research. I am therefore confident that this dissertation provides a valuable contribution towards advanced system-layer modeling of heterogeneous cellular networks, which enables to sharply predict the behavior of a realistic deployment while keeping the amount of adjustable parameters at a minimum.

Appendix A.

List of Abbreviations

3GPP	3rd Generation Partnership Project
AMC	Adaptive Modulation and Coding
AWGN	Additive White Gaussian Noise
BLER	Block-Error Ratio
BICM	Bit Interleaved Coded Modulation
BS	Base Station
CCDF	Complementary Cumulative Distribution Function
CDF	Cumulative Distribution Function
CF	Characteristic Function
CLT	Central Limit Theorem
CoMP	Coordinated Multi-Point
CSG	Closed Subscriber Group
ECDF	Empirical Cumulative Distribution Function
eICIC	Enhanced Intercell Interference Coordination
GIG	Generalized Integer Gamma
HARQ	Hybrid Automatic Repeat Request
i.i.d.	independent and identically distributed
IEEE	Institute of Electrical and Electronics Engineers
KS	Kolmogorov-Smirnov
LOS	Line of Sight
LT	Laplace Transformation
LTE	Long Term Evolution
LTE-A	LTE-Advanced
MCS	Modulation and Coding Scheme
MMSE	Minimum Mean Square Error
MGF	Moment Generating Function
MIESM	Mutual Information based exponential SNR Mapping

Appendix A. List of Abbreviations

MIMO	Multiple-Input-Multiple-Output
MISO	Multiple-Input-Single-Output
MLE	Maximum Likelihood Estimation
MMSE	Minimum Mean Square Error
MRC	Maximum Ratio Combining
MRT	Maximum Ratio Transmission
MU	Multi-User
NLOS	Non Line of Sight
OSG	Open Subscriber Group
PDF	Probability Density Function
PGFL	Probability Generating Functional
PP	Point Process
PPP	Poisson Point Process
QAM	Quadrature Amplitude Modulation
ROP	Random Object Process
RV	Random Variable
SIMO	Single-Input-Multiple-Output
SINR	Signal-to-Interference plus Noise Ratio
SIR	Signal-to-Interference Ratio
SISO	Single-Input-Single-Output
SNR	Signal-to-Noise Ratio
TM	Transmission Mode
TTI	Transmission Time Interval
UE	User Equipment
UMTS	Universal Mobile Telecommunications System
WiMAX	Worldwide Interoperability for Microwave Access
ZF	Zero Forcing

Appendix B.

Limitation of the Gaussian Approximation for the Aggregate Interference

Referring to [134], this chapter scrutinizes the validity of the Gaussian approximation for the aggregate interference in a Poisson field of interferers. The investigations are based on the Berry-Esseen bound. In its basic form, it formulates as follows.

Theorem B.1 (Berry-Esseen Theorem). *Let $\{G_i\}$ denote independent RVs with a common CDF $F(x)$. Let the RVs have zero mean, non-zero variance (i.e., $\sigma^2 > 0$) and finite third absolute moment (i.e., $\rho = \mathbb{E}[|G_i|^3] < \infty$). Then, for all x and k*

$$\|F_k(x) - F_N(x)\| \leq \frac{3\rho}{\sigma^3 \sqrt{k}}, \tag{B.1}$$

where $F_k(\cdot)$ is the CDF of the normalized sum $(1/\sigma\sqrt{k}) \sum_{i=1}^{k} G_i$, and $F_N(\cdot)$ is the CDF of the standard normal distribution, i.e., $\mathcal{N}(0, 1)$.

Proof. The proof is found in [137, p. 543]. □

It is deduced that the Gaussian approximation of the aggregate interference in a Poisson field of

Appendix B. Limitation of the Gaussian Approximation for the Aggregate Interference

interferers is valid if

$$\sqrt{\lambda \pi r_o^2} \gg 2.21 \frac{2(\alpha-1)^{3/2}}{3\alpha-2} \frac{\tilde{\mu}_3(G)}{(\tilde{\mu}_2(G))^{3/2}}, \qquad \text{for } r_o > r_c, \tag{B.2}$$

$$\sqrt{\lambda \pi r_c^2} \gg 2.21 \frac{3(\alpha-1)^{3/2}}{\sqrt{\alpha}(3\alpha-2)} \frac{\tilde{\mu}_3(G)}{(\tilde{\mu}_2(G))^{3/2}}, \qquad \text{for } r_o = 0, \tag{B.3}$$

where r_o is the inner radius of the exclusion region around the receiver of interest, r_c denotes the critical distance of the path loss law $\ell(r) = \min(b_\text{P}, 1/c_\text{P} r^{-\alpha})$, i.e., $r_c = (b_\text{P} c_\text{P})^{-1/\alpha}$, α is the path loss exponent, with $\alpha > 2$ and λ refers to the intensity of the PPP. The term $\tilde{\mu}_m(G)$ denotes the m-th *raw* moment of the RV G that models the composite fading, i.e., $\tilde{\mu}_m(G) = \mathbb{E}[G^m]$.

In (B.2), it is observed that the convergence towards Gaussianity is determined by the number of dominant interferers around the receiver of interest, and not the total number of interferers in the field. The amount of dominant interferers either grows with increasing λ or increasing r_o. As these nodes accumulate, the aggregate interference distribution converges to a Gaussian distribution by virtue of the CLT[1]. On the other hand, it is seen from (B.3) that the validity of the approximation is disputable, when there is no exclusion region at all (i.e., $r_o = 0$) or when the region is small, unless the intensity λ of the active nodes is very high, which might be infeasible in practice.

This thesis places particular emphasis on modeling the composite fading by a Gamma RV, i.e., $G \sim \Gamma[k, \theta]$. In this case,

$$\frac{\tilde{\mu}_3(G)}{(\tilde{\mu}_2(G))^{3/2}} = \frac{k+2}{\sqrt{k(k+1)}} \tag{B.4}$$

in (B.2) and (B.3), respectively. For $1/2 < k < \infty$, this expression is bounded as

$$1 < \frac{k+2}{\sqrt{k(k+1)}} < 2.89. \tag{B.5}$$

In this work, the highest intensities are specified as $\lambda = 10^{-4}\,\text{m}^{-2}$ at an exclusion radius of $r_o = 80\,\text{m}$ (conf. Chapter 3) and $\lambda = 0.5 \cdot 10^{-5}\,\text{m}^{-2}$ at $r_o = 500\,\text{m}$ (conf. Chapter 4), yielding 1.42 and 1.98 for the left-hand side of (B.2), respectively. Evaluating the right-hand side of (B.2) with the lower bound from (B.5) and $\alpha = 2$ (which, according to [134], achieves the best convergence to a Gaussian distribution) gives 1.11. Since these values clearly violate the inequality in (B.2), it can safely be assumed that the scenarios in this thesis diverge from Gaussianity.

---

[1] For further information, see, e.g., [243].

Appendix C.

Proof of Theorem 4.1

Let $G_l \sim \Gamma[k_l, \theta_l]$ be L independent Gamma RVs with $k_l \in \mathbb{N}^+$ and all θ_l different. Then, the PDF of $Y = G_1 + \cdots + G_L$ can be expressed as

$$f_Y(y) = \sum_{l=1}^{L} \frac{\Lambda_l}{\theta_l^{k_l}} h_{k_l-1,l}(0) e^{-y/\theta_l} \tag{C.1}$$

with

$$\Lambda_l = \frac{(-1)^{k_l+1}}{(k_l-1)!} \prod_{i=1, i \neq l}^{L} \left(1 - \frac{\theta_i}{\theta_l}\right)^{-k_i}, \qquad l = 1, \ldots, L \tag{C.2}$$

$$h_{\delta+1,l}(\zeta) = h_{1,l}(\zeta) h_{\delta,l}(\zeta) + \frac{d}{d\zeta} h_{\delta,l}(\zeta), \qquad \delta = 0, \ldots, k_l - 1 \tag{C.3}$$

and

$$h_{1,l}(0) = -y + \sum_{i=1, i \neq l}^{L} k_i \left(\frac{1}{\theta_i} - \frac{1}{\theta_l}\right)^{-1}, \qquad l = 1, \ldots, L \tag{C.4}$$

$$h_{1,l}^{(m)}(0) = m! \sum_{i=1, i \neq l}^{L} k_i \left(\frac{1}{\theta_i} - \frac{1}{\theta_l}\right)^{-m-1}, \qquad m = 1, \ldots, k_l - 1 \tag{C.5}$$

Appendix C. Proof of Theorem 4.1

Let $G_l \sim \Gamma[k_l, \theta_l]$ be L independent random variables with k_l being positive integers and all θ_l different. Then, the PDF of $Y = \sum_{l=1}^{L} G_l$ can be expressed as [92]

$$f_Y(y) = \left(\prod_{i=1}^{L} \frac{1}{\theta_i^{k_i}}\right) \frac{1}{2\pi i} \oint_C \frac{\prod_{i=1}^{L} \left\{\Gamma\left(\frac{1}{\theta_i} + s\right)\right\}^{k_i}}{\prod_{i=1}^{L} \left\{\Gamma\left(1 + \frac{1}{\theta_i} + s\right)\right\}^{k_i}} e^{sy} ds \qquad (C.6)$$

$$= \left(\prod_{i=1}^{L} \frac{1}{\theta_i^{k_i}}\right) \mathcal{G}_{\mathcal{K},\mathcal{K}}^{\mathcal{K},0}\left[e^{-y} \left| \begin{array}{c} \Theta_a \\ \Theta_b \end{array}\right.\right], \qquad (C.7)$$

where $\mathcal{G}_{p,q}^{m,n}[\cdot]$ denotes Meijer's G function, $\mathcal{K} = \sum_{i=1}^{L} k_i$, and

$$\Theta_a = \left\{\overbrace{\left(1+\frac{1}{\theta_1}\right),\ldots,\left(1+\frac{1}{\theta_1}\right)}^{k_1 \text{ times}},\ldots,\overbrace{\left(1+\frac{1}{\theta_L}\right),\ldots,\left(1+\frac{1}{\theta_L}\right)}^{k_L \text{ times}}\right\}, \qquad (C.8)$$

$$\Theta_b = \left\{\overbrace{\left(\frac{1}{\theta_1}\right),\ldots,\left(\frac{1}{\theta_1}\right)}^{k_1 \text{ times}},\ldots,\overbrace{\left(\frac{1}{\theta_L}\right),\ldots,\left(\frac{1}{\theta_L}\right)}^{k_L \text{ times}}\right\}. \qquad (C.9)$$

The unique values of Θ_a and Θ_b and their multiplicities k_i are gathered by the vectors **a**, **b** and **k**, respectively. Then, $|\mathbf{a}| = |\mathbf{b}| = L$, $a_i = (1 + 1/\theta_i)$ and $b_i = (1/\theta_i)$ for $i = 1, \ldots, L$.

By virtue of the calculus of residues, (C.6) can be evaluated by a summation over the negative residues of the integrand

$$I(s) = \frac{\prod_{i=1}^{L} \left\{\Gamma\left(\frac{1}{\theta_i} + s\right)\right\}^{k_i}}{\prod_{i=1}^{L} \left\{\Gamma\left(1 + \frac{1}{\theta_i} + s\right)\right\}^{k_i}} z^s \qquad (C.10)$$

as

$$\mathcal{G}_{\mathcal{K},\mathcal{K}}^{\mathcal{K},0}\left[z \left| \begin{array}{c} \Theta_a \\ \Theta_b \end{array}\right.\right] = -\sum_{l=1}^{L} \sum_{j=0}^{\infty} R_l(j). \qquad (C.11)$$

With

$$R_l(j) = \frac{1}{(k_l-1)!} \frac{d^{k_l-1}}{ds^{k_l-1}} \left\{\left(s - \left(\frac{1}{\theta_l} + j\right)\right)^{k_l} I(s)\right\}\bigg|_{s=\frac{1}{\theta_l}+j} \qquad (C.12)$$

and the substitution $s = \frac{1}{\theta_l} + k + \zeta$, it is obtained

$$R_l(j) = \frac{1}{(k_l-1)!} \frac{d^{k_l-1}}{d\zeta^{k_l-1}} g(\zeta;j) \tag{C.13}$$

$$= g_l(0;j) \frac{h_{k_l-1}(0;j)}{(k_l-1)!}. \tag{C.14}$$

Auxiliary function $g_l(0;j)$ is calculated as

$$g_l(0;j) = (-1)^{k_l} z^{1/\theta_l} \frac{\prod_{i=1, i\neq l}^{L} \Gamma(\beta_i)^{k_i}}{\prod_{i=1}^{L} \Gamma(\alpha_i)^{k_i}} \frac{z^j}{j!} \frac{\prod_{i=1}^{L} ((1-\alpha_i)_j)^{k_i}}{\prod_{i=1, i\neq l}^{L} ((1-\beta_i)_j)^{k_i}}, \tag{C.15}$$

where $(\cdot)_c$ refers to the Pochhammer symbol, which is specified as $(x)_j = x(x+1)\ldots(x+j-1)$. The therms α_i and β_i are defined as $\alpha_i = a_i - b_l$ and $\beta_i = b_i - b_l$, respectively.

Auxiliary function $h_{\delta,l}(0;j)$ is recursively determined as

$$h_{\delta+1,l}(\zeta;j) = h_{1,l}(\zeta;j) h_{\delta,l}(\zeta;j) + \frac{d}{d\zeta} h_{\delta,l}(\zeta;j). \tag{C.16}$$

It is left to provide the expressions for $h_{1,l}(\zeta;j)$ and $h_{1,l}^{(m)}(\zeta;j)$ at $\zeta = 0$:

$$h_{1,l}(0;j) = \log(z) - k_l \psi(1+j) - \sum_{i=k_l+1}^{K} \psi(\beta_i - j) + \sum_{i=1}^{K} \psi(\alpha_i - j), \tag{C.17}$$

$$h_{1,l}^{(m)}(0;j) = \frac{d^m}{d\zeta^m} h_{1,l}(\zeta;j) \bigg|_{\zeta=0} = k_l \psi^{(m)}(1) - k_l \psi^{(m)}(1+j) + $$
$$(-1)^m \left(-k_l \psi^{(m)}(1) - \sum_{i=1, i\neq l}^{L} \psi^{(m)}(\beta_i - j) + \sum_{i=1}^{L} \psi^{(m)}(\alpha_i - j) \right), \tag{C.18}$$

where $\psi^{(m)}(z) = \frac{d^m}{dz^m} \log(\Gamma(z))$ refers to the polygamma function of order m.

Since $\alpha_i = 1$ for $i = l$, the argument $(\alpha_i - j)$ in (C.17) and (C.18) can take on non-positive integer values for $j > 1$, where the polygamma function has poles of order $m+1$. These poles are however compensated by the zeros $(1-\alpha_i)$ in (C.15) due to the following facts: (i) By definition,

Appendix C. Proof of Theorem 4.1

$(0)_c = 0$ for $j \geq 1$, (ii) the zeros are of order k_l and (iii) for any non-positive integer q

$$\lim_{x \to q} (x-q)^{k_l} \psi^{(k_l-2)}(x-q) = 0, \tag{C.19}$$

$$\lim_{x \to q} (x-q)^{k_l} \left(\psi^{(0)}(x-q) \right)^{k_l-1} = 0. \tag{C.20}$$

The derivation order $(k_l - 2)$ and the exponent $k_l - 1$ in (C.19) and (C.20) correspond to the respective maximum values in $h_{k_l-1,l}(0;j)$. Consequently, $R_l(j) = 0$ for $j > 0$ and, therefore, (C.11) is simplified as

$$\mathcal{G}_{\mathcal{K},\mathcal{K}}^{\mathcal{K},0}\left[z \,\middle|\, \begin{matrix} \Theta_a \\ \Theta_b \end{matrix} \right] = -\sum_{l=1}^{L} R_l(0). \tag{C.21}$$

$R_l(0)$ is composed of $h_{\delta,l}(0;0)$ and $g_l(0;0)$.

From (C.17) and (C.18) it holds that

$$h_{1,l}(0;0) = \log(z) + \sum_{i=1, i \neq l}^{L} \frac{1}{\beta_i}, \tag{C.22}$$

$$h_{1,l}^{(m)}(0;0) = m! \sum_{i=1, i \neq l}^{L} \left(\frac{1}{\beta_i} \right)^{m+1}, \tag{C.23}$$

where the recurrence relation of the polygamma function is applied. Simple manipulations yield (4.7) and (4.8). With (4.6), $h_{\delta,l}(0;0)$ can be derived. Note that in (4.6)–(4.8) the second "0" in the argument, which stems from $j = 0$, is omitted for readability.

Considering that $\alpha_i = 1 + \beta_i$ and using the recurrence relation $\Gamma(z+1) = z\Gamma(z)$ of the Gamma function, (C.15) can be simplified as

$$g(0;0) = (-1)^{k_l} z^{1/\theta_l} \prod_{i=1, i \neq l}^{L} \left(\frac{1}{\beta_i} \right)^{k_i}. \tag{C.24}$$

Finally, (4.4) and (4.5) are obtained from (C.22)–(C.24).

Appendix D.

Mathematica® Code for Theorem 4.1

Pre-calculation of auxiliary functions

```
tableDepth = 15; (* Length of function series *)
hpt[ζ_, t_] := Piecewise[{
        {1, t == 0},
        {h1p[ζ], t == 1},
        {hpt[ζ, t - 1] h1p[ζ] + D[hpt[ζ, t - 1], ζ], t > 1}
}];
(* Compute functions in parallel *)
pH = ParallelTable[Simplify[hpt[ζ, i]], {i, 0, tableDepth}];
```

Store auxiliary functions

```
Export["AuxiliaryFunctions.mx", pH];
```

Load pre-calculated auxiliary functions

```
pH = Import["AuxiliaryFunctions.mx"];
```

Appendix D. Mathematica® Code for Theorem 4.1

PDF of sum of Gamma RVs

```
PDFSGN[ktVector_, z_] := Module[{},
    (* INPUT: "ktVector" is a list which contains all tuples {k_{cn},θ'_{cn}} *)
    (* e.g., ktVector  =   {{1,1/2},{1,1/4},{2,1/3},...}   *)
    gktVector = GatherBy[ktVector, 2]; (*Gather by duplicate theta values*)
    tU = DeleteDuplicates[Flatten[gktVector [[;;,   ;;,   2]]]]; (* Unique theta values *)
    kU = Total /@ gktVector [[;;,   ;;,   1]]; (* Accumulated k values of unique theta values *)
    L = Length[tU]; (*number of unique theta values*)

    (*Generate function  series  of helper  functions *)
    Λ[λ_Integer] := (-1)^{kU[[λ]] + 1}/(kU[[λ]] - 1)! Product[(1 -  tU[[i ]]/tU[[λ]])^{-kU[[i]]},
    {i, Complement[Range[1, L], {λ}]}];

    (* Evaluate Terms from auxiliary  functions *)
    (* NOTE h–function index 0  refers  to table  index 1 !! *)
    nsumPdfComponents = Evaluate@Table[
        tU[[λh]]^(-kU[[λh]]) Λ[λh] pH[[kU[[λh]]]] Exp[-y/tU[[λh]]]
    /.{ h1p[ζ] → -y + Sum[kU[[i]] (1/tU[[i]] - 1/tU[[λh]])^{-1},
        {i, Complement[Range[1, L],{λh}]}],
        Derivative[τ_][h1p][ζ] → τ! Sum[kU[[i]] (1/tU[[i]] - 1/tU[[λh]])^{-τ-1},
        {i,Complement[Range[1, L], {λh}]}]},
       {λh, 1, L}
    ];
    sumPdf = Plus @@ nsumPdfComponents;
    sumPdf /.  y → z
]
```

Appendix E.

Proof of Theorem 5.1

Consider a user at distance r, $0 < r \leq R_{\mathrm{I}}$, away from the center of a small cell-occupied building. Then, its coverage probability is determined as

$$P_{c,\mathrm{S}}(\delta|r) = \mathbb{P}\left[\gamma_{\mathrm{S}}(r) > \delta|r\right] = e^{-2\pi(\mu_{\mathrm{S}} I_{\mathrm{S}} + \mu_{\mathrm{M}} I_{\mathrm{M}})}, \tag{E.1}$$

where

$$I_{\mathrm{S}} = \int_{2R_{\mathrm{I}}}^{\infty} \left(\frac{\delta L_{\mathrm{W}}^{2} \ell_{\mathrm{O}}(t) e^{-(\beta_{\mathrm{B}} t + p_{\mathrm{B}})}}{\ell_{\mathrm{I}}(r) + \delta L_{\mathrm{W}}^{2} \ell_{\mathrm{O}}(t)}\right) t \, dt, \tag{E.2}$$

$$I_{\mathrm{M}} = \int_{R_{\mathrm{I}}}^{\infty} \left(1 - \frac{\frac{P_{\mathrm{S}}}{P_{\mathrm{M}}} \ell_{\mathrm{I}}(r)}{\frac{P_{\mathrm{S}}}{P_{\mathrm{M}}} \ell_{\mathrm{I}}(r) + \delta L_{\mathrm{W}} \ell(t)}\right) t \, dt. \tag{E.3}$$

Appendix E. Proof of Theorem 5.1

Applying (5.6), it follows from Campbell's theorem (see, e.g., [26]) that

$$P_{c,S}(\delta|r) = \mathbb{P}\left[\gamma_S(r) > \delta|r\right]$$

$$\stackrel{(a)}{=} \mathbb{E}\left[\exp\left(-\frac{\delta}{P_S \ell_I(r)} \sum_{i:X_i \in \Phi_M \setminus \mathcal{B}(0,R_I)} P_M G_i L_W \ell(R_i)\right)\right] \cdot$$

$$\mathbb{E}\left[\exp\left(-\frac{\delta}{P_S \ell_I(r)} \sum_{j:X_j \in \Phi_S \setminus \mathcal{B}(0,2R_I)} S_j P_S G_j L_W^2 \ell_O(R_j)\right)\right]$$

$$= \mathbb{E}\left[\prod_{i:X_i \in \Phi_M \setminus \mathcal{B}(0,R_I)} \exp\left(-\frac{\delta}{\frac{P_S}{P_M}\ell_I(r)} G_i L_W \ell(R_i)\right)\right] \cdot$$

$$\mathbb{E}\left[\prod_{j:X_j \in \Phi_S \setminus \mathcal{B}(0,2R_I)} \exp\left(-\frac{\delta}{\ell_I(r)} S_j G_j L_W^2 \ell_O(R_j)\right)\right]$$

$$\stackrel{(b)}{=} \mathbb{E}_{\Phi_M}\left[\prod_{i:X_i \in \Phi_M \setminus \mathcal{B}(0,R_I)} \frac{\frac{P_S}{P_M}\ell_I(r)}{\frac{P_S}{P_M}\ell_I(r) + \delta L_W \ell(R_i)}\right] \cdot$$

$$\mathbb{E}_{\Phi_S}\left[\prod_{j:X_j \in \Phi_S \setminus \mathcal{B}(0,2R_I)} \left(1 - \frac{\delta L_W^2 \ell_O(R_j) e^{-(\beta_B R_j + p_B)}}{\ell_I(r) + \delta L_W^2 \ell_O(R_j)}\right)\right], \quad \text{(E.4)}$$

where (a) exploits the fact that G_i are i.i.d. exponential RVs and (b) results from S_j being Bernoulli RVs with parameters $\exp(-\beta_B R_j - p_B)$. Finally, (5.7) is obtained by applying the PGFL, as explained in Section 2.1.

Appendix F.

Proof of Theorem 5.3

Consider an indoor user at distance r, $0 < r \leq R_I$ away from the center of a typical building with a small cell. Then, its coverage probability is calculated as

$$P_{c,S}(\delta|r) = \mathbb{P}\left[\gamma_S(r) > \delta|r\right] = e^{-2\pi(\mu_S I_S + \mu_M(I_L + I_N))}, \tag{F.1}$$

with $\gamma_S(\cdot)$ from (5.26) and

$$I_S = \int_{2R_I}^{\infty} \frac{\delta L_W \ell_L(t) e^{-(\beta_B t + p_B)}}{\ell_I(r) + \delta L_W \ell_L(t)} t\, dt, \tag{F.2}$$

$$I_L = \int_{R_I}^{\infty} \left(1 - \frac{\frac{P_S}{P_M}\ell_I(r)}{\frac{P_S}{P_M}\ell_I(r) + \delta \ell_L(t)}\right) tv(t)\, dt, \tag{F.3}$$

$$I_N = \int_{R_I}^{\infty} \left(1 - \frac{\frac{P_S}{P_M}\ell_I(r)}{\frac{P_S}{P_M}\ell_I(r) + \delta \ell_N(t)}\right) t(1 - v(t))\, dt. \tag{F.4}$$

Appendix F. Proof of Theorem 5.3

The proof extends the procedure in Appendix E by distinguishing LOS- and NLOS macro BSs. Again, Campbell's theorem is applied. It follows from (5.26) that

$$P_{c,S}(\delta|r) = \mathbb{P}\left[\gamma_S(r) > \delta|r\right]$$

$$= \mathbb{E}\left[\prod_{i:X_i \in \Phi_L \backslash \mathcal{B}(0,R_I)} \exp\left(-\frac{\delta}{P_S \ell_I(r)} P_M G_i \ell_L(R_i)\right)\right] \cdot$$

$$\mathbb{E}\left[\prod_{j:X_j \in \Phi_N \backslash \mathcal{B}(0,R_I)} \exp\left(-\frac{\delta}{P_S \ell_I(r)} P_M G_j \ell_N(R_j)\right)\right] \cdot$$

$$\mathbb{E}\left[\prod_{k:X_k \in \Phi_S \backslash \mathcal{B}(0,2R_I)} \exp\left(-\frac{\delta}{\ell_I(r)} S_k G_k L_W \ell_L(R_k)\right)\right]$$

$$= \mathbb{E}_{\Phi_L}\left[\prod_{i:X_i \in \Phi_L \backslash \mathcal{B}(0,R_I)} \frac{\frac{P_S}{P_M}\ell_I(r)}{\frac{P_S}{P_M}\ell_I(r) + \delta \ell_L(R_i)}\right] \cdot$$

$$\mathbb{E}_{\Phi_N}\left[\prod_{j:X_j \in \Phi_N \backslash \mathcal{B}(0,R_I)} \frac{\frac{P_S}{P_M}\ell_I(r)}{\frac{P_S}{P_M}\ell_I(r) + \delta \ell_N(R_j)}\right] \cdot$$

$$\mathbb{E}_{\Phi_S}\left[\prod_{k:X_k \in \Phi_S \backslash \mathcal{B}(0,2R_I)} 1 - \frac{\delta L_W \ell_L(R_k) e^{-(\beta_B R_k + p_B)}}{\ell_I(r) + \delta L_W \ell_L(R_k)}\right]$$

$$= \exp\left(-2\pi \mu_M \int_{R_I}^{\infty} \left(1 - \frac{\frac{P_S}{P_M}\ell_I(r)}{\frac{P_S}{P_M}\ell_I(r) + \delta \ell_L(t)}\right) t v(t) dt\right) \cdot$$

$$\exp\left(-2\pi \mu_M \int_{R_I}^{\infty} \left(1 - \frac{\frac{P_S}{P_M}\ell_I(r)}{\frac{P_S}{P_M}\ell_I(r) + \delta \ell_N(t)}\right) t (1 - v(t)) dt\right) \cdot$$

$$\exp\left(-2\pi \mu_S \int_{2R_I}^{\infty} \frac{\delta L_W \ell_L(t) e^{-(\beta_B t + p_B)}}{\ell_I(r) + \delta L_W \ell_L(t)} t dt\right), \tag{F.5}$$

where Φ_L and Φ_N denote the PPPs, which correspond to LOS- and NLOS macro BSs, and Φ_S refers to the PPP of the small cells, respectively. The according densities are given as $\mu_M v(t)$, $\mu_M(1 - v(t))$ and μ_S, respectively.

Bibliography

[1] Ericsson, "Ericsson mobility report," Aug. 2014. [Online]. Available: http://www.ericsson.com/mobility-report

[2] Cisco, "Cisco visual networking index: Global mobile data traffic forecast update, 2013-2018," Feb. 2014. [Online]. Available: http://www.cisco.com/c/en/us/solutions/collateral/service-provider/visual-networking-index-vni/white_paper_c11-520862.pdf

[3] Qualcomm, "1000x data challenge," Nov. 2013. [Online]. Available: https://www.qualcomm.com/1000x

[4] W. Webb, *Wireless Communications: The Future*. Wiley, 2007.

[5] M.-S. Alouini and A. Goldsmith, "Area spectral efficiency of cellular mobile radio systems," *IEEE Trans. Veh. Technol.*, vol. 48, no. 4, pp. 1047–1066, July 1999.

[6] V. Chandrasekhar, J. Andrews, and A. Gatherer, "Femtocell networks: A survey," *IEEE Commun. Mag.*, vol. 46, no. 9, pp. 59 –67, Sep. 2008.

[7] J. Andrews, "Seven ways that HetNets are a cellular paradigm shift," *IEEE Commun. Mag.*, vol. 51, no. 3, pp. 136–144, Mar. 2013.

[8] J. Andrews, H. Claussen, M. Dohler, S. Rangan, and M. Reed, "Femtocells: Past, present, and future," *IEEE J. Sel. Areas Commun.*, vol. 30, no. 3, pp. 497–508, Apr. 2012.

[9] A. Mesodiakaki, F. Adelantado, L. Alonso, and C. Verikoukis, "Energy-efficient context-aware user association for outdoor small cell heterogeneous networks," in *IEEE Int. Conf. Commun. (ICC)*, Sydney, NSW, Australia, June 2014, pp. 1614–1619.

[10] A. Ghosh, N. Mangalvedhe, R. Ratasuk, B. Mondal, M. Cudak, E. Visotsky, T. A. Thomas, J. G. Andrews, P. Xia, H. S. Jo *et al.*, "Heterogeneous cellular networks: From theory to practice," *IEEE Commun. Mag.*, vol. 50, no. 6, pp. 54–64, June 2012.

[11] P. Lin, J. Zhang, Y. Chen, and Q. Zhang, "Macro-femto heterogeneous network deployment and management: from business models to technical solutions," *IEEE Wireless Commun.*, vol. 18, no. 3, pp. 64–70, June 2011.

[12] 3rd Generation Partnership Project (3GPP), "Service requirements for Home Node B (HNB) and Home eNode B (HeNB)," 3rd Generation Partnership Project (3GPP), TS 22.220, Oct. 2014.

BIBLIOGRAPHY

[13] *IEEE Standard for Local and metropolitan area networks - Part 16: Air Interface for Fixed Broadband Wireless Access Systems - Amendment 2: Medium Access Control Modifications and Additional Physical Layer Specifications for 2-11 GHz*, IEEE Std., 2003.

[14] H. ElSawy, E. Hossain, and M. Haenggi, "Stochastic geometry for modeling, analysis, and design of multi-tier and cognitive cellular wireless networks: A survey," *IEEE Commun. Surveys & Tutorials*, vol. 15, no. 3, pp. 996–1019, June 2013.

[15] L. T. W. Ho and H. Claussen, "Effects of user-deployed, co-channel femtocells on the call drop probability in a residential scenario," in *IEEE Int. Symp. Personal, Indoor and Mobile Radio Commun. (PIMRC)*, Athens, Greece, Sep. 2007.

[16] M. Win, P. Pinto, and L. Shepp, "A mathematical theory of network interference and its applications," *Proc. IEEE*, vol. 97, no. 2, pp. 205–230, Feb. 2009.

[17] R. W. Heath Jr., M. Kountouris, and T. Bai, "Modeling heterogeneous network interference using Poisson point processes," *IEEE Trans. Signal Process.*, vol. 61, no. 16, pp. 4114–4126, Aug. 2013.

[18] R. W. Heath Jr. and M. Kountouris, "Modeling heterogeneous network interference," in *Inf. Theory and Appl. Workshop (ITA)*, San Diego, CA, USA, Feb. 2012, pp. 17–22.

[19] M. Haenggi and R. K. Ganti, *Interference in Large Wireless Networks*, ser. Foundations and Trends in Networking. NoW Publishers, Feb. 2009, vol. 3.

[20] A. Zemlianov and G. de Veciana, "Cooperation and decision-making in a wireless multi-provider setting," in *Joint Conf. IEEE Comput. and Commun. Soc.*, vol. 1, Mar. 2005, pp. 386–397 vol. 1.

[21] J. Andrews, F. Baccelli, and R. Ganti, "A tractable approach to coverage and rate in cellular networks," *IEEE Trans. Commun.*, vol. 59, no. 11, pp. 3122–3134, Nov. 2011.

[22] F. Baccelli, M. Klein, M. Lebourges, and S. Zuyev, "Stochastic geometry and architecture of communication networks," *Telecommun. Syst.*, vol. 7, pp. 209–227, June 1997.

[23] T. Brown, "Cellular performance bounds via shotgun cellular systems," *IEEE J. Sel. Areas Commun.*, vol. 18, no. 11, pp. 2443–2455, Nov. 2000.

[24] T. Vu, L. Decreusefond, and P. Martins, "An analytical model for evaluating outage and handover probability of cellular wireless networks," *Int. Symp. Wireless Personal Multimedia Commun. (WPMC)*, pp. 643–647, Sep. 2012.

[25] P. Madhusudhanan, J. G. Restrepo, Y. Liu, and T. X. Brown, "Carrier to interference ratio analysis for the shotgun cellular system," in *IEEE Global Telecommun. Conf. (GLOBECOM)*, Honolulu, HI, USA, Nov. 2009.

[26] F. Baccelli and B. Blaszczyszyn, *Stochastic Geometry and Wireless Networks: Volume I Theory*, ser. Foundation and Trends in Networking. Now Publishers, Mar. 2009.

BIBLIOGRAPHY

[27] A. Viterbi and I. Jacobs, "Advances in coding and modulation for noncoherent channels affected by fading, partial band, and multiple-access interference," in *Advances in Communication Systems: Theory and Applications*, vol. 4. New York, Academic Press, Inc., 1975, pp. 279–308.

[28] I. M. I. Habbab, M. Kavehrad, and C.-E. W. Sundberg, "ALOHA with capture over slow and fast fading radio channels with coding and diversity," *IEEE J. Sel. Areas Commun.*, vol. 7, no. 1, pp. 79–88, Jan. 1989.

[29] J.-P. Linnartz, H. Goossen, R. Hekmat, K. Pahlavan, and K. Zhang, "Comment on slotted ALOHA radio networks with PSK modulation in Rayleigh fading channels (and reply)," *Electron. Lett.*, vol. 26, no. 9, pp. 593–595, Apr. 1990.

[30] N. C. Beaulieu and A. A. Abu-Dayya, "Bandwidth efficient QPSK in cochannel interference and fading," *IEEE Trans. Commun.*, vol. 43, no. 9, pp. 2464–2474, Sep. 1995.

[31] J. Cheng and N. C. Beaulieu, "Accurate DS-CDMA bit-error probability calculation in Rayleigh fading," *IEEE Trans. Wireless Commun.*, vol. 1, no. 1, pp. 3–15, Jan. 2002.

[32] S. Kassam and J. Thomas, *Signal Detection in Non-Gaussian Noise*, ser. Springer texts in electrical engineering. Springer-Verlag, 1988.

[33] E. Wegman, S. Schwartz, and J. Thomas, *Topics in Non-Gaussian Signal Processing*. Springer New York, 2011.

[34] M. Chiani, "Analytical distribution of linearly modulated cochannel interferers," *IEEE Trans. Commun.*, vol. 45, no. 1, pp. 73–79, Jan. 1997.

[35] A. Giorgetti and M. Chiani, "Influence of fading on the Gaussian approximation for BPSK and QPSK with asynchronous cochannel interference," *IEEE Trans. Wireless Commun.*, vol. 4, no. 2, pp. 384–389, Mar. 2005.

[36] M. Aljuaid and H. Yanikomeroglu, "Investigating the validity of the Gaussian approximation for the distribution of the aggregate interference power in large wireless networks," in *Biennial Symp. Commun. (QBSC)*, Kingston, ON, Canada, May 2010, pp. 122–125.

[37] E. Sousa, "Performance of a spread spectrum packet radio network link in a Poisson field of interferers," *IEEE Trans. Inf. Theory*, vol. 38, no. 6, pp. 1743–1754, Nov. 1992.

[38] D. Middleton, "Non-Gaussian noise models in signal processing for telecommunications: New methods and results for class A and class B noise models," *IEEE Trans. Inf. Theory*, pp. 1129–1149, May 1999.

[39] H. Inaltekin, M. Chiang, H. V. Poor, and S. B. Wicker, "On unbounded path-loss models: Effects of singularity on wireless network performance," *IEEE J. Sel. Areas Commun.*, vol. 27, no. 7, pp. 1078–1092, Sep. 2009.

BIBLIOGRAPHY

[40] A. Rabbachin, T. Q. Quek, H. Shin, and M. Z. Win, "Cognitive network interference," *IEEE J. Sel. Areas Commun.*, vol. 29, no. 2, pp. 480–493, Feb. 2011.

[41] P. Carr, H. Geman, D. B. Madan, and M. Yor, "The fine structure of asset returns: An empirical investigation," *J. Bus.*, vol. 75, no. 2, pp. 305–333, 2002.

[42] M. Derakhshani and T. Le-Ngoc, "Aggregate interference and capacity-outage analysis in a cognitive radio network," *IEEE Trans. Veh. Technol.*, vol. 61, no. 1, pp. 196–207, Nov. 2012.

[43] J. Andrews, R. Ganti, M. Haenggi, N. Jindal, and S. Weber, "A primer on spatial modeling and analysis in wireless networks," *IEEE Commun. Mag.*, vol. 48, no. 11, pp. 156–163, Nov. 2010.

[44] G. Falciasecca, C. Caini, M. Frullone, G. Riva, and A. Serra, "Evaluation of spectral efficiency of high capacity mobile radio systems for different scenarios," in *IEEE Veh. Technol. Conf. (VTC)*, St. Louis, MO, USA, May 1991, pp. 704–709.

[45] K. Gulati, B. L. Evans, J. G. Andrews, and K. R. Tinsley, "Statistics of co-channel interference in a field of Poisson and Poisson-Poisson clustered interferers," *IEEE Trans. Signal Process.*, vol. 58, no. 12, pp. 6207–6222, Sep. 2010.

[46] M. Kountouris and N. Pappas, "Approximating the interference distribution in large wireless networks," in *Int. Symp. Wireless Commun. Syst. (ISWCS)*, Barcelona, Spain, Aug. 2014.

[47] M. Haenggi, J. Andrews, F. Baccelli, O. Dousse, and M. Franceschetti, "Stochastic geometry and random graphs for the analysis and design of wireless networks," *IEEE J. Sel. Areas Commun.*, vol. 27, no. 7, pp. 1029–1046, Sep. 2009.

[48] D. Avidor and S. Mukherjee, "Hidden issues in the simulation of fixed wireless systems," *Wireless Networks*, vol. 7, no. 2, pp. 187–200, Mar. 2001. [Online]. Available: http://dx.doi.org/10.1023/A:1016641723805

[49] F. Baccelli and B. Blaszczyszyn, *Stochastic Geometry and Wireless Networks, Volume II - Applications*, ser. Foundations and Trends in Networking, F. Baccelli and B. Blaszczyszyn, Eds. NoW Publishers, Mar. 2009, vol. 2.

[50] X. Yang and G. de Veciana, "Inducing multiscale clustering using multistage MAC contention in CDMA ad hoc networks," *IEEE/ACM Trans. Netw.*, vol. 15, no. 6, pp. 1387–1400, Dec. 2007.

[51] C.-S. Chiu and C.-C. Lin, "Comparative downlink shared channel evaluation of WCDMA release 99 and HSDPA," in *IEEE Int. Conf. Networking, Sensing and Control*, vol. 2, Mar. 2004, pp. 1165–1170.

[52] R. Jain, C. So-In, and A.-k. Al Tamimi, "System-level modeling of IEEE 802.16 E mobile WiMAX networks: key issues," *IEEE Wireless Commun.*, vol. 15, no. 5, pp. 73–79, Oct. 2008.

[53] J. C. Ikuno, M. Wrulich, and M. Rupp, "System level simulation of LTE networks," in *IEEE Veh. Technol. Conf. (VTC Spring)*, Taipei, Taiwan, May 2010.

BIBLIOGRAPHY

[54] J.-S. Wu, J.-K. Chung, and M.-T. Sze, "Analysis of uplink and downlink capacities for two-tier cellular system," *IEEE Proc. Commun.*, vol. 144, no. 6, pp. 405–411, Dec. 1997.

[55] R. S. Karlsson, "Radio resource sharing and capacity of some multiple access methods in hierarchical cell structures," in *IEEE Veh. Technol. Conf. (VTC Fall)*, vol. 5, Amsterdam, Netherlands, Sep. 1999, pp. 2825–2829.

[56] V. Mac Donald, "Advanced mobile phone service: The cellular concept," *The Bell Syst. Tech. J.*, vol. 58, no. 1, pp. 15–41, Jan. 1979.

[57] K. Gilhousen, I. Jacobs, R. Padovani, A. Viterbi, J. Weaver, L.A, and I. Wheatley, C.E., "On the capacity of a cellular CDMA system," *IEEE Trans. Veh. Technol.*, vol. 40, no. 2, pp. 303–312, May 1991.

[58] J. Xu, J. Zhang, and J. Andrews, "On the accuracy of the Wyner model in cellular networks," *IEEE Trans. Wireless Commun.*, vol. 10, no. 9, pp. 3098–3109, Sep. 2011.

[59] A. Ganz, C. Krishna, D. Tang, and Z. Haas, "On optimal design of multitier wireless cellular systems," *IEEE Commun. Mag.*, vol. 35, no. 2, pp. 88–93, Feb. 1997.

[60] E. Ekici and C. Ersoy, "Multi-tier cellular network dimensioning," *Wireless Networks*, vol. 7, no. 4, pp. 401–411, Jan. 2001.

[61] D. Taylor, H. Dhillon, T. Novlan, and J. Andrews, "Pairwise interaction processes for modeling cellular network topology," in *IEEE Global Commun: Conf. (GLOBECOM)*, Anaheim, CA, USA, Dec. 2012, pp. 4524–4529.

[62] V. Chandrasekhar and J. Andrews, "Uplink capacity and interference avoidance for two-tier femtocell networks," *IEEE Trans. Wireless Commun.*, vol. 8, no. 7, pp. 3498–3509, July 2009.

[63] F. Pantisano, M. Bennis, W. Saad, and M. Debbah, "Spectrum leasing as an incentive towards uplink macrocell and femtocell cooperation," *IEEE J. Sel. Areas Commun.*, vol. 30, no. 3, pp. 617–630, Apr. 2012.

[64] L. Saker, S.-E. Elayoubi, R. Combes, and T. Chahed, "Optimal control of wake up mechanisms of femtocells in heterogeneous networks," *IEEE J. Sel. Areas Commun.*, vol. 30, no. 3, pp. 664–672, Apr. 2012.

[65] C.-H. Lee, C.-Y. Shih, and Y.-S. Chen, "Stochastic geometry based models for modeling cellular networks in urban areas," *Wireless Networks*, vol. 19, no. 6, pp. 1063–1072, Aug. 2013.

[66] J. A. Mcfadden, "The entropy of a point process," *J. Soc. for Ind. and Appl. Mathematics*, vol. 13, no. 4, pp. 988–994, Dec. 1965. [Online]. Available: http://dx.doi.org/10.2307/2946418

[67] F. Baccelli and S. Zuyev, "Stochastic geometry models of mobile communication networks," in *Frontiers in queueing: models and applications in science and engineering*. CRC Press, 1996, pp. 227–243.

[68] C. Ren, J. Zhang, W. Xie, and D. Zhang, "Performance analysis for heterogeneous cellular networks based on Matern-like point process model," in *Int. Conf. Inform. Sci. and Technol. (ICIST)*, Yangzhou, China, Mar. 2013, pp. 1507–1511.

[69] A. Guo and M. Haenggi, "Spatial stochastic models and metrics for the structure of base stations in cellular networks." *IEEE Trans. Wireless Commun.*, vol. 12, no. 11, pp. 5800–5812, Oct. 2013.

[70] M. Di Renzo, A. Guidotti, and G. Corazza, "Average rate of downlink heterogeneous cellular networks over generalized fading channels: A stochastic geometry approach," *IEEE Trans. Commun.*, vol. 61, no. 7, pp. 3050–3071, July 2013.

[71] M. Haenggi, *Stochastic Geometry for Wireless Networks*, ser. Stochastic Geometry for Wireless Networks. Cambridge University Press, 2012. [Online]. Available: http://books.google.de/books?id=t8FeLgEACAAJ

[72] S. Weber and J. G. Andrews, "Transmission capacity of wireless networks," *Foundations and Trends in Networking*, vol. 5, no. 2-3, pp. 109–281, 2012. [Online]. Available: http://dx.doi.org/10.1561/1300000032

[73] T. Bai, R. Vaze, and R. W. Heath Jr., "Analysis of blockage effects on urban cellular networks," *IEEE Trans. Wireless Commun.*, vol. 13, no. 9, pp. 5070–5083, Sep. 2014.

[74] F. Baccelli, B. Blaszczyszyn, and P. Muhlethaler, "Stochastic analysis of spatial and opportunistic ALOHA," *IEEE J. Sel. Areas Commun.*, vol. 27, no. 7, pp. 1105–1119, Sep. 2009.

[75] J. Riihijarvi and P. Mahonen, "A model based approach for estimating aggregate interference in wireless networks," in *Int. ICST Conf. Cognitive Radio Oriented Wireless Networks and Commun. (CROWNCOM)*, Stockholm, Sweden, June 2012, pp. 180–184.

[76] M. Taranetz and M. Rupp, "A circular interference model for wireless cellular networks," in *Proc. Int. Wireless Commun. & Mobile Computing Conf.*, Nicosia, Cyprus, Aug. 2014.

[77] M. Taranetz and M. Rupp, "A circular interference model for asymmetric aggregate interference," 2015, submitted to IEEE Trans. Wireless Commun. [Online]. Available: http://arxiv.org/abs/1505.05842

[78] M. Taranetz, M. Rupp, R. W. Heath Jr., and T. Bai, "Analysis of small cell partitioning in urban two-tier heterogeneous cellular networks," in *Proc. Int. Symp. Wireless Commun. Syst. (ISWCS'14)*, Barcelona, Spain, Aug. 2014.

[79] M. Taranetz and M. Rupp, "Performance of femtocell access point deployments in user hot-spot scenarios," in *Australasian Telecommun. Networks and Appl. Conf.*, Brisbane, Australia, Nov. 2012.

[80] M. Taranetz, J. C. Ikuno, and M. Rupp, "Sensitivity of OFDMA-based macrocellular LTE networks to femtocell deployment density and isolation," in *Int. Symp. Wireless Commun. Syst.*, Ilmenau, Germany, Aug. 2013.

BIBLIOGRAPHY

[81] M. Taranetz, T. Blazek, T. Kropfreiter, M. K. Müller, S. Schwarz, and M. Rupp, "Runtime precoding: Enabling multipoint transmission in LTE-Advanced system level simulations," *IEEE Access*, 2015.

[82] S. Schwarz, J. C. Ikuno, M. Simko, M. Taranetz, Q. Wang, and M. Rupp, "Pushing the limits of LTE: A survey on research enhancing the standard," *IEEE Access*, vol. 1, pp. 51–62, May 2013.

[83] S. Schwarz, J. C. Ikuno, M. Simko, M. Taranetz, Q. Wang, and M. Rupp, "LTE research using the Vienna LTE link and system level simulators," in *COST IC1004 Manage. Committee and Scientific Meeting*, Bristol, United Kingdom, Sep. 2012.

[84] M. Müller, M. Taranetz, and M. Rupp, "Performance of remote unit collaboration schemes in high speed train scenarios," in *IEEE Veh. Technol. Conf. (VTC)*, Boston, MA, USA, Sep. 2015.

[85] M. Müller, M. Taranetz, and M. Rupp, "Providing current and future cellular services to high speed trains," *CoRR*, vol. abs/1505.04557, 2015, submitted to IEEE Commun. Mag. [Online]. Available: http://arxiv.org/abs/1505.04557

[86] K. Gulati, A. Chopra, B. L. Evans, and K. R. Tinsley, "Statistical modeling of co-channel interference," in *IEEE Global Telecommun. Conf. (GLOBECOM)*, Honolulu, HI, USA, Nov. 2009.

[87] A. Wyner, "Shannon-theoretic approach to a Gaussian cellular multiple-access channel," *IEEE Trans. Inf. Theory*, vol. 40, no. 6, pp. 1713–1727, Nov. 1994.

[88] J. Gertner, *The Idea Factory: Bell Labs and the Great Age of American Innovation*. Penguin Group, 2012.

[89] M.-S. Alouini, A. Abdi, and M. Kaveh, "Sum of Gamma variates and performance of wireless communication systems over Nakagami-fading channels," *IEEE Trans. Veh. Technol.*, vol. 50, no. 6, pp. 1471–1480, Nov. 2001.

[90] G. Efthymoglou and V. Aalo, "Performance of rake receivers in Nakagami fading channel with arbitrary fading parameters," *Electronics Lett.*, vol. 31, no. 18, pp. 1610–1612, Aug. 1995.

[91] V. Aalo, T. Piboongungon, and G. Efthymoglou, "Another look at the performance of MRC schemes in Nakagami-m fading channels with arbitrary parameters," *IEEE Trans. Commun.*, vol. 53, no. 12, pp. 2002–2005, Dec. 2005.

[92] I. Ansari, F. Yilmaz, M.-S. Alouini, and O. Kucur, "On the sum of Gamma random variates with application to the performance of Maximal Ratio Combining over Nakagami-m fading channels," in *IEEE Int. Workshop Signal Process. Advances in Wireless Commun. (SPAWC)*, Cesme, Turkey, June 2012, pp. 394–398.

[93] T. A. Tsiftsis, G. K. Karagiannidis, S. A. Kotsopoulos, and N. C. Sagias, "Performance of MRC diversity receivers over correlated Nakagami-m fading channels," in *Int. Symp. Commun. Syst., Networks and Digital Signal Process. (CSNDSP)*, Patras, Greece, July 2006.

[94] E. K. Al-Hussaini and A. Al-Bassiouni, "Performance of MRC diversity systems for the detection of signals with Nakagami fading," *IEEE Trans. Commun.*, vol. 33, no. 12, pp. 1315–1319, Dec. 1985.

[95] G. Karagiannidis, N. Sagias, and T. Tsiftsis, "Closed-form statistics for the sum of squared Nakagami-m variates and its applications," *IEEE Trans. Commun.*, vol. 54, no. 8, pp. 1353–1359, Aug. 2006.

[96] G. Efthymoglou, T. Piboongungon, and V. Aalo, "Performance of DS-CDMA receivers with MRC in Nakagami-m fading channels with arbitrary fading parameters," *IEEE Trans. Veh. Technol.*, vol. 55, no. 1, pp. 104–114, Jan. 2006.

[97] P. Lombardo, G. Fedele, and M. Rao, "MRC performance for binary signals in Nakagami fading with general branch correlation," *IEEE Trans. Commun.*, vol. 47, no. 1, pp. 44–52, Jan. 1999.

[98] Q. Zhang, "Exact analysis of postdetection combining for DPSK and NFSK systems over arbitrarily correlated Nakagami channels," *IEEE Trans. Commun.*, vol. 46, no. 11, pp. 1459–1467, Nov. 1998.

[99] L.-L. Yang and H.-H. Chen, "Error probability of digital communications using relay diversity over Nakagami-m fading channels," *IEEE Trans. Wireless Commun.*, vol. 7, no. 5, pp. 1806–1811, May 2008.

[100] D. J. Torrieri and M. C. Valenti, "The outage probability of a finite ad hoc network in Nakagami fading," *IEEE Trans. Commun.*, vol. 60, no. 11, pp. 3509–3518, Nov. 2012.

[101] P. Moschopoulos, "The distribution of the sum of independent Gamma random variables," *Ann. Inst. Statist. Mathematics*, vol. 37, no. 1, pp. 541–544, 1985.

[102] D. G. Kabe, "On the exact distribution of a class of multivariate test criteria," *Ann. Math. Statist.*, vol. 33, no. 3, pp. 1197–1200, Sep. 1962.

[103] E. Scheuer, "Reliability of an m-out-of-n system when component failure induces higher failure rates in survivors," *IEEE Trans. Rel.*, vol. 37, no. 1, pp. 73–74, Apr. 1988.

[104] S. Amari and R. Misra, "Closed-form expressions for distribution of sum of exponential random variables," *IEEE Trans. Rel.*, vol. 46, no. 4, pp. 519–522, Dec. 1997.

[105] C. A. Coelho, "The generalized integer Gamma distribution - a basis for distributions in multivariate statistics," *J. Multivariate Anal.*, vol. 64, no. 1, pp. 86 – 102, Jan. 1998.

[106] A. A. Abu-Dayya and N. C. Beaulieu, "Outage probabilities in the presence of correlated lognormal interferers," *IEEE Trans. Veh. Technol.*, vol. 43, no. 1, pp. 164–173, Feb. 1994.

[107] N. C. Beaulieu, A. A. Abu-Dayya, and P. J. McLane, "Estimating the distribution of a sum of independent lognormal random variables," *IEEE Trans. Commun.*, vol. 43, no. 12, pp. 2869–2873, Dec. 1995.

[108] J. Hu and N. C. Beaulieu, "Accurate simple closed-form approximations to Rayleigh sum distributions and densities," *IEEE Commun. Lett.*, vol. 9, no. 2, pp. 109–111, Feb. 2005.

[109] N. B. Mehta, J. Wu, A. F. Molisch, and J. Zhang, "Approximating a sum of random variables with a lognormal," *IEEE Trans. Wireless Commun.*, vol. 6, no. 7, pp. 2690–2699, July 2007.

[110] S. O. Rice, "Mathematical analysis of random noise," *Bell Syst. Tech. J.*, vol. 23, no. 3, pp. 282–332, July 1944.

[111] E. J. G. Pitman and E. J. Williams, *Studies in probability and statistics : Papers in honour of Edwin J. G. Pitman*. North-Holland Pub. Co, 1976.

[112] N. Campbell, "Discontinuities in light emission," *Proc. Cambridge Philosoph. Soc., Math. and Physical Sci.*, vol. 15, pp. 310–328, 1909.

[113] N. Campbell, "The study of discontinuous phenomena," *Proc. Cambridge Philosoph. Soc., Math. and Physical Sci.*, vol. 15, pp. 117–136, 1909.

[114] W. Schottky, "Über spontane Stromschwankungen in verschiedenen Elektrizitätsleitern," *Annalen der Physik*, vol. 362, no. 23, pp. 541–567, 1918. [Online]. Available: http://dx.doi.org/10.1002/andp.19183622304

[115] B. Blaszczyszyn, "What geometry for wireless networks - when honeycomb is as Poisson and what if both are not ideal," in *Int. Symp. Modeling and Optimization in Mobile, Ad Hoc and Wireless Networks (WiOpt)*, Paderborn, Germany, May 2012, pp. 330–330.

[116] N. Miyoshi, T. Shirai *et al.*, "A cellular network model with Ginibre configured base stations," *Advances in Appl. Probability*, vol. 46, no. 3, pp. 832–845, Sep. 2014.

[117] E. S. Sousa and J. A. Silvester, "Optimum transmission ranges in a direct-sequence spread-spectrum multihop packet radio network," *IEEE J. Sel. Areas Commun.*, vol. 8, no. 5, pp. 762–771, June 1990.

[118] H. Dhillon, R. Ganti, F. Baccelli, and J. Andrews, "Modeling and analysis of K-tier downlink heterogeneous cellular networks," *IEEE J. Sel. Areas Commun.*, vol. 30, no. 3, pp. 550–560, Apr. 2012.

[119] H.-S. Jo, Y. J. Sang, P. Xia, and J. Andrews, "Heterogeneous cellular networks with flexible cell association: A comprehensive downlink SINR analysis," *IEEE Trans. Wireless Commun.*, vol. 11, no. 10, pp. 3484–3495, Aug. 2012.

[120] S. Mukherjee, "Distribution of downlink SINR in heterogeneous cellular networks," *IEEE J. Sel. Areas Commun.*, vol. 30, no. 3, pp. 575–585, Apr. 2012.

[121] W. C. Cheung, T. Q. Quek, and M. Kountouris, "Throughput optimization, spectrum allocation, and access control in two-tier femtocell networks," *IEEE J. Sel. Areas Commun.*, vol. 30, no. 3, pp. 561–574, Apr. 2012.

BIBLIOGRAPHY

[122] H. ElSawy and E. Hossain, "Two-tier HetNets with cognitive femtocells: Downlink performance modeling and analysis in a multichannel environment," *IEEE Trans. Mobile Comput.*, vol. 13, no. 3, pp. 649–663, Mar. 2014.

[123] H. ElSawy and E. Hossain, "On cognitive small cells in two-tier heterogeneous networks," in *Int. Symp. Modeling Optimization in Mobile, Ad Hoc Wireless Networks*, Tsukuba Science City, Japan, May 2013, pp. 75–82.

[124] C.-H. Lee and M. Haenggi, "Interference and outage in Poisson cognitive networks," *IEEE Trans. Wireless Commun.*, vol. 11, no. 4, pp. 1392–1401, Apr. 2012.

[125] V. Chandrasekhar and J. Andrews, "Spectrum allocation in tiered cellular networks," *IEEE Trans. Commun.*, vol. 57, no. 10, pp. 3059–3068, Oct. 2009.

[126] P. Brémaud, *Mathematical principles of signal processing: Fourier and wavelet analysis*. Springer, 2002.

[127] V. Chandrasekhar and J. Andrews, "Uplink capacity and interference avoidance for two-tier cellular networks," in *IEEE Global Telecommun. Conf.*, Washington, DC, USA, Nov. 2007, pp. 3322–3326.

[128] P. Pinto, A. Giorgetti, M. Win, and M. Chiani, "A stochastic geometry approach to coexistence in heterogeneous wireless networks," *IEEE J. Sel. Areas Commun.*, vol. 27, no. 7, pp. 1268–1282, Sep. 2009.

[129] G. Alfano, M. Garetto, and E. Leonardi, "New insights into the stochastic geometry analysis of dense CSMA networks," in *IEEE Int. Conf. Comput. Commun. (INFOCOM)*, Shanghai, China, Apr. 2011, pp. 2642–2650.

[130] R. Ganti and M. Haenggi, "Interference in ad hoc networks with general motion-invariant node distributions," in *IEEE Int. Symp. Inf. Theory (ISIT)*, Toronto, ON, Canada, July 2008.

[131] Y. Wen, S. Loyka, and A. Yongacoglu, "On distribution of aggregate interference in cognitive radio networks," in *Symp. Commun. (QBSC)*, Kingston, ON, Canada, May 2010, pp. 265–268.

[132] M. Aljuaid and H. Yanikomeroglu, "A cumulant-based characterization of the aggregate interference power in wireless networks," in *IEEE Veh. Technol. Conf. (VTC)*, Taipei, Taiwan, May 2010.

[133] P. Dmochowski, P. Smith, M. Shafi, J. Andrews, and R. Mehta, "Interference models for heterogenous sources," in *IEEE Int. Conf. Commun. (ICC)*, Ottawa, ON, Canada, June 2012, pp. 4049–4054.

[134] M. Aljuaid and H. Yanikomeroglu, "Investigating the Gaussian convergence of the distribution of the aggregate interference power in large wireless networks," *IEEE Trans. Veh. Technol.*, vol. 59, no. 9, pp. 4418–4424, Aug. 2010.

BIBLIOGRAPHY

[135] A. C. Berry, "The accuracy of the Gaussian approximation to the sum of independent variates," *Trans. Amer. Mathem. Soc.*, vol. 49, no. 1, pp. pp. 122–136, 1941.

[136] R. Ganti and J. Andrews, "Correlation of link outages in low-mobility spatial wireless networks," in *Asilomar Conf. Signals, Syst. and Comput. (Asilomar)*, Pacific Grove, CA, USA, Nov. 2010, pp. 312–316.

[137] W. Feller, *An Introduction to Probability Theory and Its Applications.* Wiley, Jan. 1968, vol. 1.

[138] M. Kim, Y. Han, Y. Yoon, Y.-J. Chong, and H. Lee, "Modeling of adjacent channel interference in heterogeneous wireless networks," *IEEE Commun. Lett.*, vol. 17, no. 9, pp. 1774–1777, Sep. 2013.

[139] K. Read, "A lognormal approximation for the collector's problem," *Amer. Statistician*, vol. 52, no. 2, pp. 175–180, May 1998.

[140] A. Ghasemi and E. Sousa, "Interference aggregation in spectrum-sensing cognitive wireless networks," *IEEE J. Sel. Topics Signal Process.*, vol. 2, no. 1, pp. 41–56, Feb. 2008.

[141] R. Menon, R. M. Buehrer, and J. Reed, "On the impact of dynamic spectrum sharing techniques on legacy radio systems," *IEEE Trans. Wireless Commun.*, vol. 7, no. 11, pp. 4198–4207, Nov. 2008.

[142] M. Pratesi, F. Santucci, and F. Graziosi, "Generalized moment matching for the linear combination of lognormal RVs - application to outage analysis in wireless systems," in *IEEE Int. Symp. Personal, Indoor and Mobile Radio Commun. (PIMRC)*, Helsinki, Finland, Sep. 2006.

[143] A. Abdi and M. Kaveh, "On the utility of Gamma PDF in modeling shadow fading (slow fading)," in *IEEE Veh. Technol. Conf. (VTC)*, vol. 3, Houston, TX, USA, July 1999, pp. 2308–2312.

[144] R. W. Heath Jr., T. Wu, Y. H. Kwon, and A. Soong, "Multiuser MIMO in distributed antenna systems with out-of-cell interference," *IEEE Trans. Signal Process.*, vol. 59, no. 10, pp. 4885–4899, July 2011.

[145] S. Kusaladharma and C. Tellambura, "Aggregate interference analysis for underlay cognitive radio networks," *IEEE Wireless Commun. Lett.*, vol. 1, no. 6, pp. 641–644, Dec. 2012.

[146] I. Kostic, "Analytical approach to performance analysis for channel subject to shadowing and fading," *IEEE Proc. Commun.*, vol. 152, pp. 821–827(6), Dec. 2005.

[147] J. Zhang, M. Matthaiou, Z. Tan, and H. Wang, "Performance analysis of digital communication systems over composite $\eta - \mu$ Gamma fading channels," *IEEE Trans. Veh. Technol.*, vol. 61, no. 7, pp. 3114–3124, Sep. 2012.

[148] S. Al-Ahmadi and H. Yanikomeroglu, "On the approximation of the generalized-K PDF by a Gamma PDF using the moment matching method," in *IEEE Wireless Commun. and Networking Conf. (WCNC)*, Budapest, Hungary, Apr. 2009.

BIBLIOGRAPHY

[149] S. Al-Ahmadi and H. Yanikomeroglu, "On the use of high-order moment matching to approximate the generalized-K distribution by a Gamma distribution," in *IEEE Global Telecommun. Conf. (GLOBECOM)*, Honolulu, HI, USA, Nov. 2009.

[150] M. G. Kendall et al., "The advanced theory of statistics." *The Advanced Theory of Statist.*, no. 2nd Ed, 1946.

[151] C. H. de Lima, M. Bennis, and M. Latva-aho, "Coordination mechanisms for self-organizing femtocells in two-tier coexistence scenarios," *IEEE Trans. Wireless Commun.*, vol. 11, no. 6, pp. 2212–2223, Apr. 2012.

[152] A. Hyvärinen, J. Karhunen, and E. Oja, *Independent component analysis*. John Wiley & Sons, 2004, vol. 46.

[153] C. C. Chan and S. Hanly, "Calculating the outage probability in a CDMA network with spatial Poisson traffic," *IEEE Trans. Veh. Technol.*, vol. 50, no. 1, pp. 183–204, Jan. 2001.

[154] R. Menon, R. M. Buehrer, and J. H. Reed, "Outage probability based comparison of underlay and overlay spectrum sharing techniques," in *IEEE Int. Symp. New Frontiers in Dynamic Spectrum Access Networks*, Baltimore, MD, USA, Nov. 2005, pp. 101–109.

[155] M. Laner, P. Svoboda, and M. Rupp, "Parsimonious network traffic modeling by transformed ARMA models," *IEEE Access*, vol. 2, pp. 40–55, Jan. 2014.

[156] T. M. Cover and J. A. Thomas, *Elements of information theory*. John Wiley & Sons, 2012.

[157] C. A. Coelho and J. T. Mexia, "On the distribution of the product and ratio of independent generalized Gamma-ratio random variables," *Sankhyā: The Indian J. Statist.*, vol. 69, no. 12, pp. 221–255, May 2007.

[158] P. Marsch and G. Fettweis, "Static clustering for Cooperative Multi-Point (CoMP) in mobile communications," in *IEEE Int. Conf. Commun. (ICC)*, Kyoto, Japan, June 2011.

[159] C. Ball, R. Mullner, J. Lienhart, and H. Winkler, "Performance analysis of closed and open loop MIMO in LTE," in *European Wireless Conf. (EW)*, Aalborg, Denmark, May 2009, pp. 260–265.

[160] A. Farajidana, W. Chen, A. Damnjanovic, T. Yoo, D. Malladi, and C. Lott, "3GPP LTE downlink system performance," in *IEEE Global Telecommun. Conf. (GLOBECOM)*, Honolulu, HI, USA, Dec. 2009.

[161] J. Giese, M. Amin, and S. Brueck, "Application of coordinated beam selection in heterogeneous LTE-Advanced networks," in *Int. Symp. Wireless Commun. Syst. (ISWCS)*, Aachen, Germany, Nov. 2011, pp. 730–734.

[162] Y. Liang, A. Goldsmith, G. Foschini, R. Valenzuela, and D. Chizhik, "Evolution of base stations in cellular networks: Denser deployment versus coordination," in *IEEE Int. Conf. Commun. (ICC)*, Beijing, China, May 2008, pp. 4128–4132.

BIBLIOGRAPHY

[163] T. Novlan, R. Ganti, and J. Andrews, "Coverage in two-tier cellular networks with fractional frequency reuse," in *IEEE Global Telecommun. Conf. (GLOBECOM)*, Houston, TX, USA, Dec. 2011.

[164] F. Di Salvo, "A characterization of the distribution of a weighted sum of Gamma variables through multiple hypergeometric functions," *Integral Transforms and Special Functions*, vol. 19, no. 8, pp. 563–575, Aug. 2008.

[165] Y. Zhuang, Y. Luo, L. Cai, and J. Pan, "A geometric probability model for capacity analysis and interference estimation in wireless mobile cellular systems," in *IEEE Global Telecommun. Conf. (GLOBECOM)*, Houston, TX, USA, Dec. 2011.

[166] K. B. Baltzis, *Cellular Networks - Positioning, Performance Analysis, Reliability*. InTech, Apr. 2011, ch. Hexagonal vs Circular Cell Shape: A Comparative Analysis and Evaluation of the Two Popular Modeling Approximations.

[167] 3rd Generation Partnership Project (3GPP), "Evolved universal terrestrial radio access (E-UTRA); radio frequency (RF) system scenarios," 3rd Generation Partnership Project (3GPP), TR 36.942, Oct. 2014.

[168] I. S. Gradshteyn and I. M. Ryzhik, *Table of Integrals, Series, and Products*, 7th ed. Elsevier/Academic Press, Amsterdam, Feb. 2007.

[169] N. Bhushan, J. Li, D. Malladi, R. Gilmore, D. Brenner, A. Damnjanovic, R. Sukhavasi, C. Patel, and S. Geirhofer, "Network densification: the dominant theme for wireless evolution into 5G," *IEEE Commun. Mag.*, vol. 52, no. 2, pp. 82–89, Feb. 2014.

[170] H. Tabassum, F. Yilmaz, Z. Dawy, and M.-S. Alouini, "A framework for uplink intercell interference modeling with channel-based scheduling," *IEEE Trans. Wireless Commun.*, vol. 12, no. 1, pp. 206–217, Jan. 2013.

[171] R. Prasad and A. Kegel, "Improved assessment of interference limits in cellular radio performance," *IEEE Trans. Veh. Technol.*, vol. 40, no. 2, pp. 412–419, May 1991.

[172] S. B. P. Iman Mabrouk, "The exact density function of a sum of independent Gamma random variables as an inverse Mellin transform," *Int. J. Appl. Math. and Statist.*, vol. 41, no. 11, 2013.

[173] S. B. Provost, "On sums of independent Gamma random variates," *Statist.: J. Theoretical and Appl. Statist.*, vol. 20, no. 4, 1989.

[174] S. Plass, X. G. Doukopoulos, and R. Legouable, "Investigations on link-level inter-cell interference in OFDMA systems," in *Symp. Commun. and Veh. Technol. (SCVT)*, Liege, Belgium, Nov. 2006, pp. 49–52.

[175] 3rd Generation Partnership Project (3GPP), "Evolved Universal Terrestrial Radio Access (E-UTRA); mobility enhancements in heterogeneous networks," 3rd Generation Partnership Project (3GPP), TR 36.839, Jan. 2013.

BIBLIOGRAPHY

[176] 3rd Generation Partnership Project (3GPP), "Coordinated multi-point operation for LTE physical layer aspects," 3rd Generation Partnership Project (3GPP), TR 36.819, Sep. 2013.

[177] J. H. Curtiss, "On the distribution of the quotient of two chance variables," *Ann Math. Statist.*, vol. 12, no. 4, pp. 409–421, Dec. 1941.

[178] A. Lozano, R. W. Heath Jr., and J. Andrews, "Fundamental limits of cooperation," *IEEE Trans. Inf. Theory*, vol. 59, no. 9, pp. 5213–5226, Sep. 2013.

[179] P. Xia, V. Chandrasekhar, and J. G. Andrews, "Open vs. closed access femtocells in the uplink," *IEEE Trans. Wireless Commun.*, vol. 9, no. 12, pp. 3798–3809, Oct. 2010.

[180] H. Wang and M. Reed, "Tractable model for heterogeneous cellular networks with directional antennas," in *Australian Commun. Theory Workshop*, Wellington, Australia, Jan. 2012, pp. 61–65.

[181] S. Mukherjee, "UE coverage in LTE macro network with mixed CSG and open access femto overlay," in *IEEE Int. Conf. Commun. Workshops (ICC)*, Kyoto, Japan, June 2011.

[182] V. Chandrasekhar, M. Kountouris, and J. Andrews, "Coverage in tiered cellular networks with spatial diversity," in *IEEE Global Telecommun. Conf.*, Honolulu, HI, USA, Nov. 2009.

[183] V. Chandrasekhar, J. Andrews, Z. Shen, T. Muharemovict, and A. Gatherer, "Distributed power control in femtocell-underlay cellular networks," in *IEEE Global Telecommun. Conf.*, Honolulu, HI, USA, Nov. 2009.

[184] V. Chandrasekhar, M. Kountouris, and J. G. Andrews, "Coverage in multi-antenna two-tier networks," *IEEE Trans. Wireless Commun.*, vol. 8, no. 10, pp. 5314–5327, Dec. 2014.

[185] 3rd Generation Partnership Project (3GPP), "Study on 3D channel model for LTE," 3rd Generation Partnership Project (3GPP), TR 36.873, Sep. 2014.

[186] R. Cowan, "Objects arranged randomly in space: An accessible theory," *Advances in Appl. Probability*, vol. 21, no. 3, pp. 543–569, Sep. 1989.

[187] S. Chiu, D. Stoyan, W. Kendall, and J. Mecke, *Stochastic Geometry and Its Applications*, ser. Wiley Series in Probability and Statistics. Wiley, Aug. 2013.

[188] B. Hanci and I. Cavdar, "Mobile radio propagation measurements and tuning the path loss model in urban areas at GSM-900 band in Istanbul-Turkey," in *IEEE Veh. Technol. Conf. (VTC)*, vol. 1, Sep. 2004, pp. 139–143 Vol. 1.

[189] G. Boudreau, J. Panicker, N. Guo, R. Chang, N. Wang, and S. Vrzic, "Interference coordination and cancellation for 4G networks," *IEEE Commun. Mag.*, vol. 47, no. 4, pp. 74–81, Apr. 2009.

[190] 3rd Generation Partnership Project (3GPP), "Evolved Universal Terrestrial Radio Access (E-UTRA) physical channels and modulation," 3rd Generation Partnership Project (3GPP), TS 36.211, Jan. 2015.

BIBLIOGRAPHY

[191] T. Bai and R. W. Heath Jr., "Coverage analysis for millimeter wave cellular networks with blockage effects," in *IEEE Global Conf. Signal and Inform. Process. (GlobalSIP)*, Austin, TX, USA, Dec. 2013, pp. 727–730.

[192] M. Di Renzo, "Stochastic geometry modeling and analysis of multi-tier millimeter wave cellular networks," *CoRR*, vol. abs/1410.3577, Oct. 2014.

[193] R. Gahleitner and E. Bonek, "Radio wave penetration into urban buildings in small cells and microcells," in *IEEE Veh. Technol. Conf. (VTC)*, Stockholm, Sweden, June 1994, pp. 887–891 vol.2.

[194] Y. Nagata, Y. Furuya, E. Moriyama, M. Mizuno, I. Kamiya, and S. Hattori, "Measurement and modeling of 2 GHz-band out-of-sight radio propagation characteristics under microcellular environments," in *IEEE Int. Symp. Personal, Indoor and Mobile Radio Commun. (PIMRC)*, Sydney, NSW, Australia, Sep. 1991, pp. 341–346.

[195] T. Schwengler and M. Glbert, "Propagation models at 5.8 GHz - path loss and building penetration," in *IEEE Radio and Wireless Conf. (RAWCON)*, Denver, CO, USA, Sep. 2000, pp. 119–124.

[196] N. Papadakis, A. Kanatas, and P. Constantinou, "Microcellular propagation measurements and simulation at 1.8 GHz in urban radio environment," *IEEE Trans. Veh. Techn.*, vol. 47, no. 3, pp. 1012–1026, Aug. 1998.

[197] J. Andersen, T. Rappaport, and S. Yoshida, "Propagation measurements and models for wireless communications channels," *IEEE Commun. Mag.*, vol. 33, no. 1, pp. 42–49, Jan. 1995.

[198] F. Kakar, K. Sani, and F. Elahi, "Essential factors influencing building penetration loss," in *IEEE Int. Conf. Commun. Technol. (ICCT)*, Hangzhou, China, Nov. 2008.

[199] J.-E. Berg, "Building penetration loss at 1700 MHz along line of sight street microcells," in *IEEE Int. Symp. Personal, Indoor and Mobile Radio Commun. (PIMRC)*, Boston, MA, USA, Oct. 1992, pp. 86–87.

[200] A. De Toledo and A. M. Turkmani, "Propagation into and within buildings at 900, 1800 and 2300 MHz," in *IEEE Veh. Technol. Conf. (VTC)*, Denver, CO, USA, May 1992, pp. 633–636 vol.2.

[201] H. Masui, T. Kobayashi, and M. Akaike, "Microwave path-loss modeling in urban line-of-sight environments," *IEEE J. Sel. Areas Commun.*, vol. 20, no. 6, pp. 1151–1155, Aug. 2002.

[202] 3rd Generation Partnership Project (3GPP), "Evolved Universal Terrestrial Radio Access (E-UTRA); Further advancements for E-UTRA physical layer aspects," 3rd Generation Partnership Project (3GPP), TR 36.814, Mar. 2010.

[203] T. Rappaport, *Wireless Communications: Principles and Practice*, 2nd ed. Upper Saddle River, NJ, USA: Prentice Hall PTR, 2001.

BIBLIOGRAPHY

[204] X. Zhao, J. Kivinen, P. Vainikainen, and K. Skog, "Propagation characteristics for wideband outdoor mobile communications at 5.3 GHz," *IEEE J. Sel. Areas Commun.*, vol. 20, no. 3, pp. 507–514, Apr. 2002.

[205] Y. Corre, J. Stephan, and Y. Lostanlen, "Indoor-to-outdoor path-loss models for femtocell predictions," in *IEEE Int. Symp. Personal Indoor and Mobile Radio Commun. (PIMRC)*, Toronto, ON, Canada, Sep. 2011, pp. 824–828.

[206] H. Dhillon and J. Andrews, "Downlink rate distribution in heterogeneous cellular networks under generalized cell selection," *IEEE Wireless Commun. Lett.*, vol. 3, no. 1, pp. 42–45, Feb. 2014.

[207] I. Latif, F. Kaltenberger, and R. Knopp, "Link abstraction for multi-user MIMO in LTE using interference-aware receiver," in *IEEE Wireless Commun. and Networking Conf. (WCNC)*, Shanghai, China, Apr. 2012, pp. 842–846.

[208] M. Döttling and et al., "Assessment of Advanced Beamforming and MIMO Technologies," IST2003-507581 WINNER, Tech. Rep. D2.7 ver 1.1, May 2005.

[209] Mathworks®, "MATLAB Documentation," http://www.mathworks.com/help/matlab/index.html, Mar. 2015.

[210] K. Abdallah, I. Cerutti, and P. Castoldi, "Energy-efficient coordinated sleep of LTE cells," in *IEEE Int. Conf. Commun. (ICC)*, Ottawa, ON, Canada, June 2012, pp. 5238–5242.

[211] M. Carvalho and P. Vieira, "An enhanced handover oscillation control algorithm in LTE self-optimizing networks," in *IEEE Int. Symp. Wireless Personal Multimedia Commun. (WPMC)*, Brest, France, Oct. 2011.

[212] M. M. Selim, M. El-Khamy, and M. El-Sharkawy, "Enhanced frequency reuse schemes for interference management in LTE femtocell networks," in *Int. Symp. Wireless Commun. Syst. (ISWCS)*, Paris, France, Aug. 2012, pp. 326–330.

[213] S. Y. Shin and D. Triwicaksono, "Radio resource control scheme for machine-to-machine communication in lte infrastructure," in *Int. Conf. ICT Convergence (ICTC)*, Jeju Island, South Korea, Oct. 2012.

[214] J. C. Ikuno, "System level modeling and optimization of the LTE downlink," Ph.D. dissertation, E389, Vienna University of Technology, 2013.

[215] R. W. Heath Jr., M. Airy, and A. Paulraj, "Multiuser diversity for MIMO wireless systems with linear receivers," in *Asilomar Conf. Signals, Syst. and Comput. (Asilomar)*, vol. 2, Pacific Grove, CA, USA, Nov. 2001, pp. 1194–1199.

[216] J. C. Ikuno, C. Mehlführer, and M. Rupp, "A novel link error prediction model for OFDM systems with HARQ," in *IEEE Int. Conf. Commun. (ICC)*, Kyoto, Japan, June 2011.

BIBLIOGRAPHY

[217] S. Caban, M. Rupp, C. Mehlführer, and M. Wrulich, *Evaluation of HSDPA and LTE: From Testbed Measurements to System Level Performance.* Wiley, 2011.

[218] H. Claussen, "Efficient modelling of channel maps with correlated shadow fading in mobile radio systems," in *IEEE Int. Symp. Personal, Indoor and Mobile Radio Commun. (PIMRC)*, vol. 1, Berlin, Germany, Sep. 2005, pp. 512–516.

[219] 3rd Generation Partnership Project (3GPP), "Evolved Universal Terrestrial Radio Access (E-UTRA); Physical layer procedures," 3rd Generation Partnership Project (3GPP), TS 36.213, Jan. 2015.

[220] S. Schwarz, M. Simko, and M. Rupp, "On performance bounds for MIMO OFDM based wireless communication systems," in *IEEE Int. Workshop Signal Process. Advances in Wireless Commun. (SPAWC)*, San Francisco, CA, USA, June 2011, pp. 311–315.

[221] A. Burr, A. Papadogiannis, and T. Jiang, "MIMO truncated Shannon bound for system level capacity evaluation of wireless networks," in *IEEE Wireless Commun. and Networking Conf. Workshops (WCNCW)*, Paris, France, Apr. 2012, pp. 268–272.

[222] E. Pateromichelakis, M. Shariat, A. ul Quddus, and R. Tafazolli, "On the analysis of co-tier interference in femtocells," in *IEEE Int. Symp. Personal Indoor and Mobile Radio Commun. (PIMRC)*, Toronto, ON, Canada, Sep. 2011, pp. 122–126.

[223] F. Capozzi, G. Piro, L. Grieco, G. Boggia, and P. Camarda, "On accurate simulations of LTE femtocells using an open source simulator," *EURASIP J. Wireless Commun. and Networking*, vol. 1, Oct. 2012.

[224] J. Ling, D. Chizhik, and R. Valenzuela, "On resource allocation in dense femto-deployments," in *IEEE Int. Conf. Microwaves, Commun., Antennas and Electronics Syst. (COMCAS)*, Tel Aviv, Israel, Nov. 2009.

[225] D. Calin, H. Claussen, and H. Uzunalioglu, "On femto deployment architectures and macrocell offloading benefits in joint macro-femto deployments," *IEEE Commun. Mag.*, vol. 48, no. 1, pp. 26–32, Jan. 2010.

[226] D. Lopez-Perez, A. Valcarce, G. de la Roche, E. Liu, and J. Zhang, "Access methods to WiMAX femtocells: A downlink system-level case study," in *IEEE Int. Conf. Commun. Syst. (ICCS)*, Guangzhou, China, Nov. 2008, pp. 1657–1662.

[227] A. Lawson and D. Denison, *Spatial Cluster Modelling.* Taylor & Francis, 2002. [Online]. Available: http://books.google.at/books?id=XmQkJf3DIp8C

[228] U. Schilcher, M. Gyarmati, C. Bettstetter, Y. W. Chung, and Y. H. Kim, "Measuring inhomogeneity in spatial distributions," in *Veh. Technol. Conf. (VTC Spring)*, Singapore, May 2008, pp. 2690–2694.

BIBLIOGRAPHY

[229] 3rd Generation Partnership Project (3GPP), "Evolved Universal Terrestrial Radio Access (E-UTRA); FDD Home eNode B (HeNB) Radio Frequency (RF) requirements analysis," 3rd Generation Partnership Project (3GPP), TR 36.921, Sep. 2014.

[230] S. Seidel, T. Rappaport, S. Jain, M. Lord, and R. Singh, "Path loss, scattering and multipath delay statistics in four european cities for digital cellular and microcellular radiotelephone," *IEEE Trans. Veh. Technol.*, vol. 40, no. 4, pp. 721–730, Nov. 1991.

[231] V. Erceg, L. Greenstein, S. Tjandra, S. Parkoff, A. Gupta, B. Kulic, A. Julius, and R. Bianchi, "An empirically based path loss model for wireless channels in suburban environments," *IEEE J. Sel. Areas Commun.*, vol. 17, no. 7, pp. 1205–1211, July 1999.

[232] M. J. Feuerstein, K. L. Blackard, T. S. Rappaport, S. Y. Seidel, and H. Xia, "Path loss, delay spread, and outage models as functions of antenna height for microcellular system design," *IEEE Trans. Veh. Technol.*, vol. 43, no. 3, pp. 487–498, Aug. 1994.

[233] V. Abhayawardhana, I. Wassell, D. Crosby, M. Sellars, and M. Brown, "Comparison of empirical propagation path loss models for fixed wireless access systems," in *IEEE Veh. Technol. Conf. (VTC)*, vol. 1, May 2005, pp. 73–77.

[234] G. Durgin, T. S. Rappaport, and H. Xu, "Measurements and models for radio path loss and penetration loss in and around homes and trees at 5.85 GHz," *IEEE Trans. Commun.*, vol. 46, no. 11, pp. 1484–1496, Nov. 1998.

[235] J. Porter, I. Lisica, and G. Buchwald, "Wideband mobile propagation measurements at 3.7 GHz in an urban environment," in *IEEE Antennas and Propagation Soc. Int. Symp.*, vol. 4, June 2004, pp. 3645–3648 Vol.4.

[236] T. Rautiainen, K. Kalliola, and J. Juntunen, "Wideband radio propagation characteristics at 5.3 GHz in suburban environments," *IEEE Int. Symp. Personal, Indoor, and Mobile Radio Commun. (PIMRC)*, Sep. 2005.

[237] S. Singh, H. Dhillon, and J. Andrews, "Downlink rate distribution in multi-RAT heterogeneous networks," in *IEEE Int. Conf. Communications (ICC)*, Budapest, Hungary, June 2013, pp. 5188–5193.

[238] T. Zahir, K. Arshad, Y. Ko, and K. Moessner, "A downlink power control scheme for interference avoidance in femtocells," in *Int. Wireless Commun. and Mobile Computing Conference (IWCMC)*, Istanbul, Turkey, July 2011, pp. 1222–1226.

[239] H.-S. Jo, P. Xia, and J. G. Andrews, "Open, closed, and shared access femtocells in the downlink," *EURASIP J. Wireless Commun. and Networking*, vol. 2012, no. 1, Dec. 2012. [Online]. Available: http://dx.doi.org/10.1186/1687-1499-2012-363

[240] S. Ahmadi, *LTE-Advanced: A Practical Systems Approach to Understanding 3GPP LTE Releases 10 and 11 Radio Access Technologies*, ser. ITPro collection. Elsevier Science, 2013.

[241] A. Adhikary and G. Caire, "On the coexistence of macrocell spatial multiplexing and cognitive femtocells," in *IEEE Int. Conf. Commun. (ICC)*, June 2012, pp. 6830–6834.

[242] P. M. Dixon, "Ripley's K function," in *Encyclopedia of Environmetrics*. John Wiley & Sons, Ltd, 2001. [Online]. Available: http://dx.doi.org/10.1002/9780470057339.var046

[243] L. Le Cam *et al.*, "The central limit theorem around 1935," *Statist. Sci.*, vol. 1, no. 1, pp. 78–91, 1986.

I want morebooks!

Buy your books fast and straightforward online - at one of the world's fastest growing online book stores! Environmentally sound due to Print-on-Demand technologies.

Buy your books online at
www.get-morebooks.com

Kaufen Sie Ihre Bücher schnell und unkompliziert online – auf einer der am schnellsten wachsenden Buchhandelsplattformen weltweit!
Dank Print-On-Demand umwelt- und ressourcenschonend produziert.

Bücher schneller online kaufen
www.morebooks.de

OmniScriptum Marketing DEU GmbH
Heinrich-Böcking-Str. 6-8
D - 66121 Saarbrücken
Telefax: +49 681 93 81 567-9

info@omniscriptum.com
www.omniscriptum.com

Printed by Books on Demand GmbH, Norderstedt / Germany